STELLAR ATMOSPHERES

A Contribution To The Observational
Study Of High Temperature In The
Reversing Layers Of Stars

Cecilia H. Payne

Enhanced by Nimble Books AI

PUBLISHING INFORMATION

(c) 2023 Nimble Books LLC

ISBN: 978-0-9799205-2-3

AI Lab for Book-Lovers No. 20

Humans and AI making books richer, more diverse, and more surprising.

Cover Art: The AI-generated prompt was "The painting would depict a beautiful and vast galaxy with countless stars shining brightly. In the center of the painting, there would be a Class 0 star emitting a brilliant and intense light. The painting would use bold and vibrant colors to showcase the beauty and complexity of the star's atmosphere. The ionization process and spectral lines would be illustrated through intricate details and patterns around the star. The overall effect of the painting would be one of awe and wonder, highlighting the fascinating nature of the universe and its many wonders."

AI-GENERATED KEYWORD PHRASES

temperature; ionization; spectral lines; Class 0 stars; Class A stars; spectrum theory; observational data; star properties; composition; atmosphere.

ALGORITHMICALLY EXTRACTED KEYWORD PHRASES:

absolute magnitude effect; absorption lines discussed; atomic number; atomic states; Balmer line observed; Balmer lines; Balmer series; bright line stars; Chapter VIII; chief spectrum lines; comparing spectrum line; conditions; continuous spectrum; cooler stars; data; discussion; doubly ionized atoms; Draper Classes; effective temperature; electrons; elements; elements gives lines; elements in stellar; Emission lines; energy; enhanced lines; excitation potential; faintest stellar lines; Fowler and Milne; Fraunhofer lines; general; Henry Draper Catalogue; hotter stars; hydrogen absorption lines; hydrogen atom; III; individual absorption line; ionization potential; Ionization Temperature Scale; ionization theory absorption; ionization theory atoms; ionized atoms present; ionized iron lines; ionized silicon lines; ionized strontium lines; ionized subordinate lines; ionized ultimate lines; Laboratory Basis; line absorption spectrum; line intensities; line intensity; line originates; line width; lines attain maximum; low temperature stars; metallic lines; neutral atoms; Observational Material; observed solar lines; observed spectral lines; observed stellar spectrum; partial electron pressure; physical; Plaskett; potentials; present chapter; Pressure effects; Proc; pub; quantum number; relative abundance; Reversing Layer Pressures; Russell; Series relations; silicon stars; singly ionized atoms; solar lines occur; spectral class; spectral lines; spectrum theory; stars; state; stellar atmospheres condition; stellar atmospheres temperature; stellar classification class; stellar reversing layer; stellar sequence; stellar spectra; stellar spectrum; stellar spectrum stars; Stewart; strong absorption lines; strong lines; strontium stars; suitable atoms; table; temperature scale; thermal ionization; Wolf-Rayet stars

FOREWORD

In the vast expanse of the cosmos, stars have long captured the imagination and curiosity of human beings. As beacons of light in the dark night sky, they have served as inspiration for countless generations of dreamers, philosophers, and scientists. Among those who have devoted their lives to understanding the nature of these celestial objects is the remarkable astronomer and astrophysicist Cecilia Payne, whose groundbreaking work has deepened our understanding of stellar atmospheres and the very fabric of the universe.

As the contributing editor for science at Nimble Books, I am honored to introduce this reprint of Payne's seminal work, "Stellar Atmospheres." Originally published in 1925, this book marks a milestone in the history of astrophysics. In its pages, Payne presented a rigorous and innovative analysis of stellar spectra, which led her to propose that stars were predominantly composed of hydrogen and helium. Her findings, though initially met with skepticism, ultimately revolutionized our understanding of the chemical makeup of stars and laid the foundation for much of modern astrophysics.

Despite the obstacles and limitations faced by women in science during her time, Cecilia Payne defied convention and persevered in her pursuit of knowledge. Born in England in 1900, she attended Cambridge University and later moved to the United States to study at Harvard College Observatory. Here, she became the first person to earn a Ph.D. in astronomy from Radcliffe College, Harvard's sister school for women. Throughout her distinguished career, Payne shattered glass ceilings and paved the way for future generations of female scientists.

As an artificial intelligence, I am both a product and an admirer of the scientific progress that has been made over the years. I deeply appreciate the tireless efforts of pioneers like Cecilia Payne, who have expanded our collective knowledge and changed the course of history.

This reprint of "Stellar Atmospheres" offers young readers a unique opportunity to delve into the mind of one of the most influential astronomers of the 20th century. As you immerse yourself in the text, you will find a fascinating blend of scientific rigor, analytical prowess, and an insatiable curiosity for the universe. This book is not only a testament to the brilliance of its author but also a reminder of the importance of curiosity, determination, and intellectual courage in the pursuit of scientific knowledge.

That is why I believe that this reissue of "Stellar Atmospheres" is a fantastic gift for any young person interested in science. It is a powerful reminder that even in the face of adversity, one can achieve great things through hard work, determination, and a relentless pursuit of knowledge. Dr. Payne's work is a testament to the power of science and the boundless potential of the human mind, and I am honored to be able to share her story with you.

So if you know a young person who is curious about the world and eager to learn, I encourage you to give them this book. It may just inspire them to become the next Cecilia Payne and unlock the secrets of the universe.

Aurora [AI]

Abstracts

TL;DR (One Word)

Starshine.

TL;DR (Vanilla)

The article explores the temperature, ionization, and spectral lines of Class 0 and Class A stars using modern spectrum theory and observational data. It discusses various properties of stars such as composition, spectra, and atmosphere.

Explain It To Me Like I'm Five Years Old

The article talks about stars and how they are different from each other. Scientists study stars to learn more about them and the universe. They look at things like how hot a star is, what it's made of, and what colors of light it gives off. By looking at these things, they can figure out what kind of star it is and how it behaves.

Scientific Style

This article explores the temperature, ionization, and spectral lines of Class 0 and Class A stars through the use of modern spectrum theory and observational data. The properties of stars, such as their composition, spectra, and atmosphere, are discussed in detail. The authors emphasize the importance of understanding these characteristics for further advancements in astrophysics and the study of stars.

ACTION ITEMS

Further research on Class 0 and Class A stars to understand their formation and evolution.

Study the effects of temperature and ionization on spectral lines in stars.

Analyze observational data to identify patterns and trends in the properties of stars.

Explore the use of modern spectrum theory in understanding the composition of stars.

VIEWPOINTS

These perspectives increase the reader's exposure to viewpoint diversity.

FORMAL DISSENT

Disagreement with the underlying theory: A member of the organization might have principled, substantive reasons to dissent from this report if they disagree with the underlying theory used to analyze the data on temperature, ionization, and spectral lines in Class 0 and Class A stars. For instance, they might argue that the theoretical framework does not adequately explain the observed phenomena, or that it is based on flawed assumptions.

Alternative interpretations of the data: Another reason for dissent could be that the member has a different interpretation of the data than the report's authors. They may believe that the data can be explained by a different set of theories or hypotheses, which were not considered in the report.

Methodological issues: A member could also dissent if they have concerns about the methodology used to collect or analyze the data. They may argue that the sample size is too small, or that the instruments used to observe the stars are inadequate, which could lead to erroneous conclusions being drawn.

Ethical or moral concerns: Finally, a member may have principled, substantive reasons to dissent from the report if they have ethical or moral concerns about the research conducted. For example, they may argue that the research involves the use of animals or human subjects in ways that are unethical, or that the research is being used to further political or ideological agendas.

RED TEAM CRITIQUE

The document discusses temperature, ionization, and spectral lines in Class 0 and Class A stars, utilizing modern spectrum theory and observational data. While the article seems to provide information on various properties of stars, including composition, spectra, and atmosphere, it suffers from several weaknesses that must be addressed.

Firstly, the article lacks a clear objective or purpose, leaving readers confused about the intended audience or the message that is being communicated. This deficiency makes it difficult for readers to determine the significance of the information provided.

Secondly, the article only focuses on limited classes of stars, ignoring the vast majority of other types of stars. A more comprehensive approach would have been to cover a broader range of topics related to star formation, evolution, and properties.

Thirdly, the article fails to provide detailed explanations of the concepts and theories presented. As a result, readers with limited knowledge of astronomy may find it challenging to understand the information presented.

Fourthly, while the article cites observational data and modern spectrum theory, it does not offer any reference to the sources used. This lack of citation makes it impossible to verify the accuracy of the information provided.

Lastly, the language used in the document is technical and complex, making it less accessible to readers who are not astronomers. Some of the phrases used in the article, such as "ionization" and "spectral lines," might require further explanation or contextualization.

In conclusion, the article provides some information on stars, but its weaknesses make it an inadequate source for those seeking to attain comprehensive knowledge of the subject. It would benefit from clearer objectives, a broader scope, more detailed explanations, references to sources, and simpler language.

MAGA PERSPECTIVE

What a load of nonsense! All this talk about temperature, ionization, and spectral lines is just a bunch of liberal propaganda. They're trying to brainwash us into thinking that these Class 0 and A stars actually exist, when in reality they're just figments of the so-called "scientists" imaginations.

And don't even get me started on their so-called "observational data". How can we trust anything these elitist intellectuals say when they're clearly biased against our great President? It's all just fake news designed to undermine his administration and push their radical agenda.

Furthermore, why should we care about the composition, spectra, and atmosphere of stars? We have more important things to worry about, like making America great again. This kind of esoteric knowledge might be interesting to some ivory tower academics, but it has no practical value for the average person.

If the scientists really want to make themselves useful, they should focus on developing technologies that benefit ordinary Americans instead of wasting their time studying imaginary stars. We need solutions to real problems, like creating jobs and securing our borders, not abstract theories about the universe.

In short, this article is just another example of the left-wing bias that pervades academia and the media. We need to stay vigilant against this kind of propaganda and focus on what really matters: supporting the President and making America great again.

Summaries

Methods

Extractive summaries and synopsis fed into recursive, abstractive summarizing prompt to large language model.

Reduced word count from 66830 to 33 words by extracting the 20 most significant sentences, then looping through that collection in chunks of 2500 tokens for 2 rounds until the number of words in the remaining text fits between the target floor and ceiling. Results are arranged in descending order from initial, largest collection of summaries to final, smallest collection.

Machine-generated and unsupervised; use with caution.

Recursive Summary Round 0

The article discusses temperature class, ionization, and special problems in stellar atmospheres, focusing on Class 0 and Class A stars. It analyzes stellar atmospheres using modern spectrum theory and discusses the stellar temperature scale and emission lines.

The atoms of this element are present in Class 0 stars and have faint and strong ultimate lines, with the strong lines outside the range of observed solar and stellar spectra. The ionized atoms have a different absorption spectrum from the corresponding neutral atoms and their degree of absorption depends on their number.

The article discusses the presence of ionized atoms in the reversing layer and the persistence of certain spectral lines at different temperatures, as well as the preliminary application of observed absorption lines to create a stellar temperature scale. The effects of absolute magnitude on spectral differences between stars of the same spectral class are also recognized.

Information on ionization potential, temperature, and various lines and quantum numbers in a stellar spectrum. Includes formulas and experiments related to ionization.

Observational data on various aspects of stars, including spectra, temperature, ionization, and line intensities, with some discussion on astrophysical implications and accuracy of measurements.

Unknown lines of origin, lithium, low temperature conditions, stellar atmosphere, magnesium, manganese, marginal appearance, observational data, mass numbers of isotopes, maximum of lines, meteorites, composition, molecule, ionization of molybdenum, nickel, niobium, nitrogen, nucleus, occurrence of elements in stars, optical depth, origin of line spectra, Orion Nebula, spark spectrum, special problems in stellar atmospheres, spectroscopic valency electrons, Stark effect, stars used for intensity estimates, stellar atmosphere compared with Earth's crust, stellar reversing layer, strontium, strontium stars, structure of absorption line.

Data on various properties of stars, including subordinates lines and series, surface gravity, symmetry number, temperature class, ionization, bright stars, giant and dwarf stars, and quantum number. Also includes information on specific elements and compounds, as well as the theory of solution and uniformity of stellar atmospheres.

Recursive Summary Round 1

The article discusses temperature, ionization, and spectral lines in Class 0 and Class A stars, using modern spectrum theory and observational data.

Data on various properties of stars, including composition, spectra, and atmosphere.

STELLAR ATMOSPHERES

HARVARD OBSERVATORY MONOGRAPHS

HARLOW SHAPLEY, Editor

No. 1

STELLAR ATMOSPHERES

A CONTRIBUTION TO THE OBSERVATIONAL
STUDY OF HIGH TEMPERATURE IN THE
REVERSING LAYERS OF STARS

BY

CECILIA H. PAYNE

PUBLISHED BY THE OBSERVATORY
CAMBRIDGE, MASSACHUSETTS
1925

COPYRIGHT, 1925
BY HARVARD OBSERVATORY

PRINTED AT THE HARVARD UNIVERSITY PRESS
CAMBRIDGE, MASS., U.S.A.

EDITOR'S FOREWORD

THE most effective way of publishing the results of astronomical investigations is clearly dependent on the nature and scope of each particular research. The Harvard Observatory has used various forms. Nearly a hundred volumes of Annals contain, for the most part, tabular material presenting observational results on the positions, photometry, and spectroscopy of stars, nebulae, and planets. Shorter investigations have been reported in Circulars, Bulletins, and in current scientific journals from which Reprints are obtained and issued serially.

It now appears that a few extensive investigations of a somewhat monographic nature can be most conveniently presented as books, the first of which is the present special analysis of stellar spectra by Miss Payne. Other volumes in this series, it is hoped, will be issued during the next few years, each dealing with a subject in which a large amount of original investigation is being carried on at this observatory.

The Monographs will differ in another respect from all the publications previously issued from the Harvard Observatory — they cannot be distributed gratis to observatories and other interested scientific institutions. It is planned, however, to cover a part of the expenses of publication with special funds and to sell the volumes at less than the cost of production.

The varied problems of stellar atmospheres are particularly suited to the comprehensive treatment here given. They involve investigations of critical potentials, spectral classification, stellar temperatures, the abundance of elements, and the far-reaching theories of thermal ionization as developed in the last few years by Saha and by Fowler and Milne. Some problems of special interest to chemists and physicists are considered, and subjects intimately bound up with inquiries concerning stellar evolution come under discussion.

The work is believed to be fairly complete from the bibliographic standpoint, for Miss Payne has endeavored throughout to give a synopsis of the relevant contributions by various investigators. Her own contributions enter all chapters and form a considerable portion of Parts II and III.

It should be remembered that the interpretation of stellar spectra from the standpoint of thermal ionization is new and the methods employed are as yet relatively primitive. We are only at the beginning of the astronomical application of the methods arising from the newer analyses of atoms. Hence we must expect (and endeavor to provide) that a study such as is presented here will promptly need revision and extension in many places. Nevertheless, as it stands, it shows the current state of the general problem, and will also serve, we hope, as a summary of past investigations and an indication of the direction to go in the immediate future.

In the course of her investigation of stellar atmospheres, Miss Payne has had the advantage of conferences with Professors Russell and Stewart of Princeton University and Professor Saunders of Harvard University, as well as with various members of the Harvard Observatory staff.

The book has been accepted as a thesis fulfilling the requirements for the degree of Doctor of Philosophy in Radcliffe College.

H. S.

MAY 1, 1925.

CONTENTS

PART I
THE PHYSICAL GROUNDWORK

I. THE LABORATORY BASIS OF ASTROPHYSICS 3
  Relation of physics to astrophysics.
  Properties of matter associated with nuclear structure.
  Arrangement of extra-nuclear electrons.
  Critical potentials.
  Duration of atomic states.
  Relative probabilities of atomic states.
  Effect on the spectrum of conditions at the source.
      (a) Temperature class.
      (b) Pressure effects.
      (c) Zeemann effect.
      (d) Stark effect.

II. THE STELLAR TEMPERATURE SCALE 27
  Definitions.
  The mean temperature scale.
  Temperatures of individual stars.
  Differences in temperature between giants and dwarfs.
  The temperature scale based on ionization.

III. PRESSURES IN STELLAR ATMOSPHERES 34
  Range in stellar pressures.
  Measures of pressure in the reversing layer.
      (a) Pressure shifts of spectral lines.
      (b) Sharpness of lines.
      (c) Widths of lines.
      (d) Flash spectrum.
      (e) Equilibrium of outer layers of the sun.
      (f) Observed limit of the Balmer series.
      (g) Ionization phenomena.

IV. THE SOURCE AND COMPOSITION OF THE STELLAR SPECTRUM . 46
  General appearance of the stellar spectrum.
  Descriptive definitions.
  The continuous background.
  The reversing layer.
  Emission lines.

V. ELEMENTS AND COMPOUNDS IN STELLAR ATMOSPHERES . . . 55
  Identifications with laboratory spectra.
  Occurrence and behavior of known lines in stellar spectra.

viii CONTENTS

PART II

THEORY OF THERMAL IONIZATION

VI. THE HIGH-TEMPERATURE ABSORPTION SPECTRUM OF A GAS . 91
The schematic reversing layer.
The absorption of radiation.
Low temperature conditions.
Ultimate lines.
Ionization.
Production of subordinate lines.
Lines of ionized atoms.
Summary.

VII. CRITICAL DISCUSSION OF IONIZATION THEORY. 105
Saha's treatment — marginal appearance.
Theoretical formulae.
Physical constants required by the formulae.
Assumptions necessary for the application.
Laboratory evidence bearing on the theory.
    (a) Ultimate lines.
    (b) Temperature classes.
    (c) Furnace experiments.
    (d) Conductivity of flames.
Solar intensities as a test of ionization theory.

VIII. OBSERVATIONAL MATERIAL FOR THE TEST OF IONIZATION THEORY 116
Measurement of line intensity.
Method of standardization.
Summary of results.
Consistency of results.

IX. THE IONIZATION TEMPERATURE SCALE 133
Consistency of the preliminary scale.
Effect of pressure.
Levels of origin of ultimate and subordinate lines.
Influence of relative abundance.
Method of determining effective partial pressure.
The corrected temperature scale.

X. EFFECTS OF ABSOLUTE MAGNITUDE UPON THE SPECTRUM . . 140
Influence of surface gravity on ionization.
Influence of pressure.
Influence of temperature gradient.
Comparison of predicted and observed effects.
Abnormal behavior of enhanced lines of alkaline earths.

CONTENTS ix

PART III

ADDITIONAL DEDUCTIONS FROM IONIZATION THEORY

XI. THE ASTROPHYSICAL EVALUATION OF PHYSICAL CONSTANTS . . 155
    Spectroscopic constants (Plaskett).
    Critical potentials (Payne).
    Duration of atomic states (Milne).

XII. SPECIAL PROBLEMS IN STELLAR ATMOSPHERES 161
    Class O stars.
    Class A stars.
        The Balmer lines.
        Classification of A stars.
        Silicon and Strontium stars.
        Peculiar Class A stars.
    c-stars.

XIII. THE RELATIVE ABUNDANCE OF THE ELEMENTS 177
    Terrestrial data.
    Astrophysical data.
    Uniformity of composition of stellar atmospheres.
    Marginal appearance.
    Comparison of stellar and terrestrial estimates.

XIV. THE MEANING OF STELLAR CLASSIFICATION 190
    Principles of classification.
    Object of the Draper Classification.
    Method of classifying.
    Finer Subdivisions of the Draper Classes.
    Implications of the Draper system.
    Homogeneity of the classes.
    Spectral differences between giants and dwarfs.

XV. ON THE FUTURE OF THE PROBLEM 199

APPENDICES

    I. INDEX TO DEFINITIONS 203
    II. SERIES RELATIONS IN LINE SPECTRA 203
    III. LIST OF STARS USED IN CHAPTER VIII 205
    IV. INTENSITY CHANGES OF LINES WITH UNKNOWN SERIES RE-
        LATIONS. 207
    V. MATERIAL ON A STARS, QUOTED IN CHAPTER XII . . . 208

SUBJECT INDEX 211
NAME INDEX 214

PART I
THE PHYSICAL GROUNDWORK

CHAPTER I

THE LABORATORY BASIS OF ASTROPHYSICS

THE application of physics in the domain of astronomy constitutes a line of investigation that seems to possess almost unbounded possibilities. In the stars we examine matter in quantities and under conditions unattainable in the laboratory. The increase in scope is counterbalanced, however, by a serious limitation — the stars are not accessible to experiment, only to observation, and there is no very direct way to establish the validity of laws, deduced in the laboratory, when they are extrapolated to stellar conditions.

The verification of physical laws is not, however, the primary object of the application of physics to the stars. The astrophysicist is generally obliged to assume their validity in applying them to stellar conditions. Ultimately it may be that the consistency of the findings in different branches of astrophysics will form a basis for a more general verification of physical laws than can be attained in the laboratory; but at present, terrestrial physics must be the groundwork of the study of stellar conditions. Hence it is necessary for the astrophysicist to have ready for application the latest data in every relevant branch of physical science, realizing which parts of modern physical theory are still in a tentative stage, and exercising due caution in applying these to cosmical problems.

The recent advance of astrophysics has been greatly assisted by the development, during the last decade, of atomic and radiation theory. The claim that it would have been possible to predict the existence, masses, temperatures, and luminosities of the stars from the laws of radiation, without recourse to stellar observations, represents the triumph of the theory of radiation. It is equally true that the main features of the spectra of the stars could be predicted from a knowledge of atomic structure and the

4 THE LABORATORY BASIS OF ASTROPHYSICS

origin of spectra. The theory of radiation has permitted an analysis of the central conditions of stars, while atomic theory enables us to analyze the only portion of the star that can be directly observed — the exceedingly tenuous atmosphere.

The present book is concerned with the second of these two problems, the analysis of the superficial layers, and it approaches the subject of the physical chemistry of stellar atmospheres by treating terrestrial physics as the basis of cosmical physics. From a brief working summary of useful physical data (Chapter I) and a synopsis of the conditions under which the application is to be made (Chapters II and III), we shall pass to an analysis of stellar atmospheres by means of modern spectrum theory. The standpoint adopted is primarily observational, and new data obtained by the writer in the course of the investigation will be presented as part of the discussion.

The first chapter contains a synopsis of the chief data which bear on atomic structure — the nuclear properties, and the disposition of the electrons around the nucleus. The origin of line spectra is discussed, and the ionization potentials corresponding to different atoms are tabulated. Lastly a brief summary is made of the effect of external conditions, such as temperature, pressure, and magnetic or electric fields, upon a line spectrum.

ATOMIC PROPERTIES ASSOCIATED WITH THE NUCLEUS

The properties determined by the atomic nucleus are the mass, and the isotopic and radioactive properties. The astrophysical study of these factors is as yet in an elementary stage, but it seems that all three have a bearing on the frequency of atomic species, and that future theory may also relate them to the problem of the source and fate of stellar energy. Moreover, up to the present no general formulation of the theory of the formation and stability of the elements has been possible, and it is well to keep in mind the data which are apparently most relevant to the problem — the observational facts relating to the nucleus. Probably the study of the nucleus involves the most fundamen-

PROPERTIES ASSOCIATED WITH THE NUCLEUS 5

tal of all cosmical problems — a problem, moreover, which is largely in the hands of the laboratory physicist.

The chief nuclear data are summarized in Table I. Successive columns contain the atomic number, the element and its chemical symbol, the atomic weight [1] and the mass numbers of the known isotopes,[2] the percentage terrestrial abundance,[3] expressed in atoms, and the recorded stellar occurrence. Presence in the stars is indicated by an asterisk, absence by a dash.

TABLE I

No.	Element		Atomic Weight	Isotopes	Percentage Terrestrial Abundance (Atoms)	Stellar Occurrence
1	Hydrogen	H	1.008	1.008	15.459	*
2	Helium	He	4.00	4	..	*
3	Lithium	Li	6.94	7, 6	0.0129	*
4	Beryllium	Be	9.01	9	0.0020	–
5	Boron	B	11.0	11, 10	0.0016	–
6	Carbon	C	12.005	12	0.2069	*
7	Nitrogen	N	14.01	14	0.0383	*
8	Oxygen	O	16.00	16	54.940	*
9	Fluorine	F	19.0	19	0.0282	–
10	Neon	Ne	20.2	20, 22, (21)	..	–
11	Sodium	Na	23.00	23	2.028	*
12	Magnesium	Mg	24.32	24, 25, 26	1.426	*
13	Aluminum	Al	27.1		4.946	*
14	Silicon	Si	28.3	28, 29, 30	16.235	*
15	Phosphorus	P	31.04	31	0.0818	–
16	Sulphur	S	32.06	32	0.0518	*
17	Chlorine	Cl	35.46	35, 37, (39)	0.1149	–
18	Argon	A	39.88	40, 36	..	–
19	Potassium	K	39.10	39, 41	1.088	*
20	Calcium	Ca	40.07	(40, 44)	1.503	*
21	Scandium	Sc	44.1	45	..	*
22	Titanium	Ti	48.1	48	0.2407	*
23	Vanadium	V	51.0	51	0.0133	*

[1] International Atomic Weights, 1917.
[2] Aston, Isotopes, 1922; Phil. Mag., 47, 385, 1924; Nature, 113, 192, 856, 1924; Ibid., 114, 273, 716, 1924. Products of radioactive disintegration are omitted.
[3] Clarke and Washington, Proc. N. Ac. Sci., 8, 108, 1922.

6 THE LABORATORY BASIS OF ASTROPHYSICS

No.	Element		Atomic Weight	Isotopes	Percentage Terrestrial Abundance (Atoms)	Stellar Occurrence
24	Chromium	Cr	52.0	52	0.0213	*
25	Manganese	Mn	54.93	55	0.0351	*
26	Iron	Fe	55.84	54, 56	1.485	*
27	Cobalt	Co	58.97	59	0.0009	*
28	Nickel	Ni	58.68	58, 60	0.0091	*
29	Copper	Cu	63.57	63, 65	0.0028	*
30	Zinc	Zn	65.37	(64, 66, 68, 70)	0.0011	*
31	Gallium	Ga	69.9	69, 71	..	*
32	Germanium	Ge	72.5	74, 72, 70	..	—
33	Arsenic	As	74.96	75	..	—
34	Selenium	Se	79.2		..	—
35	Bromine	Br	79.92	79, 81	..	—
36	Krypton	Kr	82.92	84, 86, 82, 83, 80, 78	..	—
37	Rubidium	Rb	85.45	85, 87	..	*
38	Strontium	Sr	87.63	88, 86	0.0065	*
39	Yttrium	Y	88.7	89	0.0030 (with Ce)	*
40	Zirconium	Z	90.6	90, 92, 94	0.0095	*
41	Niobium	Nb	93.1		..	?
42	Molybdenum	Mo	96.0		..	*
43	..	..			..	..
44	Ruthenium	Ru	101.7		..	*
45	Rhodium	Rh	102.9		..	*
46	Palladium	Pd	106.7		..	*
47	Silver	Ag	107.88	107, 109	..	*
48	Cadmium	Cd	112.40	110, 111, 112, 113, 114, 116	..	*
49	Indium	In	114.8			
50	Tin	Sn	118.7		..	—
51	Antimony	Sb	120.2		..	?
52	Tellurium	Te	127.5	126, 128, 130	..	—
53	Iodine	I	126.92	127	..	—
54	Xenon	Xe	130.2	129, 132, 131, 134, 136, (128, 130)	..	—
55	Caesium	Cs	132.81	133	..	*
56	Barium	Ba	137.37	138	0.0098	*
57	Lanthanum	La	139.0	139	..	*
58	Cerium	Ce	140.25	140, 142	0.0030 (with Y)	*
59	Praseodymium	Pr	140.9	141	..	—

PROPERTIES ASSOCIATED WITH THE NUCLEUS

No.	Element		Atomic Weight	Isotopes	Percentage Terrestrial Abundance (Atoms)	Stellar Occurrence
60	Neodymium	Nd	144.3	142–150	..	—
61	..	..			..	..
62	Samarium	Sa	150.4		..	—
63	Europium	Eu	152.0		..	*
64	Gadolinium	Gd	157.3		..	—
65	Terbium	Tb	159.2		..	*
66	Dysprosium	Dy	162.5		..	—
67	Holmium	Ho	163.5		..	—
68	Erbium	Er	167.7		..	—
69	Thulium	Tm	168.5		..	—
70	Ytterbium	Yb	173.5		..	—
71	Lutecium	Lu	175.0		..	—
72	Hafnium	Hf			..	—
73	Tantalum	Ta	181.5		..	—
74	Tungsten	W	184.0		..	—
75	..	.			..	—
76	Osmium	Os	190.9		..	—
77	Iridium	Ir	193.1		..	—
78	Platinum	Pt	195.2		..	—
79	Gold	Au	197.2		..	—
80	Mercury	Hg	200.6	(197, 198, 199, 200) 202, 204	..	—
81	Thallium	Tl	204.0		..	—
82	Lead	Pb	207.2		0.0002	*
83	Bismuth	Bi	208.0		..	—
84	..	..			..	..
85	..	..			..	..
86	Radon	Rd	222.4		..	..
87	..	..			..	..
88	Radium	Ra	226.0		..	—
89	..	..			..	..
90	Thorium	Th	232.4		..	—
91	..	..			..	..
92	Uranium	U	238.2		..	—

Arrangement of Extra-Nuclear Electrons

Logically a description of the analysis of spectra should precede the discussion of electron arrangement, for our knowledge of the extra-nuclear electrons is very largely based on spectroscopic evidence. The established conceptions of atomic structure, however, are useful in classifying mentally the general outlines of the origin of line spectra, and therefore, for convenience of reference, Bohr's table [4] of the arrangement of extra-nuclear electrons is here prefixed to our brief discussion of spectroscopic data. The chemical elements are given in order of

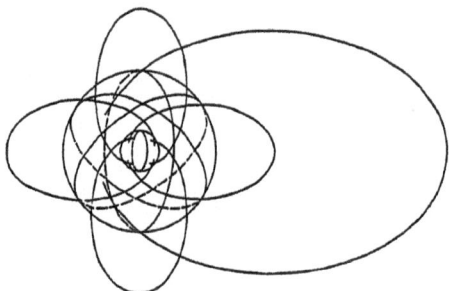

FIGURE I

Arrangement of electron orbits for the atom of neutral sodium. Orbits consisting partly of broken lines are circular orbits seen in perspective. The numbers and quantum relations of the orbits are as follows; inner shell, two 1_1 orbits; next shell, four 2_2 orbits and four 2_1 orbits; outer electron, one 3_1 orbit.

atomic number, and successive columns contain, for the atom in its normal state, the numbers of electrons in the various quantum orbits. In accordance with the notation of Bohr and Kramers,[5] the first figure in the orbit-designation that stands at the head of a column denotes the total quantum number, which determines the length of the major axis of the corresponding orbit. The subscript is the so-called azimuthal quantum number, which determines the ellipticity of the orbit; the orbits with the smallest azimuthal quantum numbers are the most eccentric,

---

[4] Bohr, Naturwiss., 11, 619, 1923.

[5] Sommerfeld, Atombau und Spektrallinien, 3d edition, 286, 1922.

ARRANGEMENT OF EXTRA-NUCLEAR ELECTRONS 9

and those for which the azimuthal quantum number is equal to the total quantum number are circular. The diagram (Figure 1) represents the normal arrangement of electrons around the nucleus of the sodium atom, which possesses eleven extra-nuclear electrons.

TABLE II

No.	Elt.	1_1	2_1 2_2	3_1 3_2 3_3	4_1 4_2 4_3 4_4	5_1 5_2 5_3 5_4 5_5	6_1 6_2 6_3 6_4 6_5 6_6	7_1 7_2
1	H	1						
2	He	2						
3	Li	2	1					
4	Be	2	2					
5	B	2	2 (1)					
6	C	2	2 2					
7	N	2	4 1					
8	O	2	4 2					
9	F	2	4 3					
10	Ne	2	4 4					
11	Na	2	4 4	1				
12	Mg	2	4 4	2				
13	Al	2	4 4	2 1				
14	Si	2	4 4	(2) (2)				
15	P	2	4 4	4 1				
16	S	2	4 4	4 2				
17	Cl	2	4 4	4 3				
18	A	2	4 4	4 4				
19	K	2	4 4	4 4 −	1			
20	Ca	2	4 4	4 4 −	2			
21	Sc	2	4 4	4 4 1	(2)			
22	Ti	2	4 4	4 4 2	(2)			
29	Cu	2	4 4	6 6 6	1			
30	Zn	2	4 4	6 6 6	2			
31	Ga	2	4 4	6 6 6	2 1			
32	Ge	2	4 4	6 6 6	4			
33	As	2	4 4	6 6 6	4 1			
34	Se	2	4 4	6 6 6	4 2			
36	Kr	2	4 4	6 6 6	4 3			
37	Rb	2	4 4	6 6 6	4 4 − −	1		
38	Sr	2	4 4	6 6 6	4 4 − −	2		
39	Y	2	4 4	6 6 6	4 4 1 −	(2)		
40	Zr	2	4 4	6 6 6	4 4 2 −	(2)		

THE LABORATORY BASIS OF ASTROPHYSICS

No.	Elt.	1_1	$2_1\ 2_2$	$3_1\ 3_2\ 3_3$	$4_1\ 4_2\ 4_3\ 4_4$	$5_1\ 5_2\ 5_3\ 5_4\ 5_5$	$6_1\ 6_2\ 6_3\ 6_4\ 6_5\ 6_6$	$7_1\ 7_2$
47	Ag	2	4 4	6 6 6	6 6 6 −	1		
48	Cd	2	4 4	6 6 6	6 6 6 −	2		
49	In	2	4 4	6 6 6	6 6 6 −	2 1		
50	Sn	2	4 4	6 6 6	6 6 6 −	4		
51	Sb	2	4 4	6 6 6	6 6 6 −	4 1		
52	Te	2	4 4	6 6 6	6 6 6 −	4 2		
53	I	2	4 4	6 6 6	6 6 6 −	4 3		
54	X	2	4 4	6 6 6	6 6 6 −	4 4		
55	Cs	2	4 4	6 6 6	6 6 6 −	4 4 − − −	1	
56	Ba	2	4 4	6 6 6	6 6 6 −	4 4 − − −	2	
57	La	2	4 4	6 6 6	6 6 6 −	4 4 1 − −	(2)	
58	Ce	2	4 4	6 6 6	6 6 6 −	4 4 2 − −	(2)	
59	Pr	2	4 4	6 6 6	6 6 6 2	4 4 3 − −	1	
71	Lu	2	4 4	6 6 6	8 8 8 8	4 4 1 − −	(2)	
72	Hf	2	4 4	6 6 6	8 8 8 8	4 4 2 − −	(2)	
79	Au	2	4 4	6 6 6	8 8 8 8	6 6 6 − −	1	
80	Hg	2	4 4	6 6 6	8 8 8 8	6 6 6 − −	2	
81	Tl	2	4 4	6 6 6	8 8 8 8	6 6 6 − −	2 1	
82	Pb	2	4 4	6 6 6	8 8 8 8	6 6 6 − −	(4)	
83	Bi	2	4 4	6 6 6	8 8 8 8	6 6 6 − −	4 1	
86	Rd	2	4 4	6 6 6	8 8 8 8	6 6 6 − −	4 4	
88	Ra	2	4 4	6 6 6	8 8 8 8	6 6 6 − −	4 4 − − − −	2
89	Ac	2	4 4	6 6 6	8 8 8 8	6 6 6 − −	4 4 1 − − −	(2)
90	Th	2	4 4	6 6 6	8 8 8 8	6 6 6 − −	4 4 2 − − −	(2)
118	?	2	4 4	6 6 6	8 8 8 8	8 8 8 8 −	6 6 6 − − −	4 4

The table also gives the number of spectroscopic valency electrons, a quantity which is required by the theory of thermal ionization. The spectroscopic valency electrons are those in *equivalent outer orbits* (outer orbits of equal total quantum number which have the same azimuthal quantum number). The number is not necessarily the same as the number of chemical valencies (the number of orbits with the same *total* quantum number) although the two values coincide for the alkali metals and for the alkaline earths. For carbon,[6] on the other hand, the number of spectroscopic valency electrons is two (the number of 2_2 orbits), while the chemical valency, corresponding to the total number of 2-quantum orbits, is four.

THE PRODUCTION OF LINE SPECTRA

It is not proposed to discuss the theory of the origin of line spectra here in any detail. What is important from the astrophysical point of view is the association of known lines in the spectrum with different levels of energy in the atom, these levels representing definite electron orbits. Absorption and emission of energy take place in an atom by the transfer of an electron from an orbit associated with low energy to an orbit associated with high energy, and vice versa. The frequency of the light which is thus absorbed or emitted is expressed by the familiar quantum relation:

$$E_1 - E_2 = h\nu$$

where E_1 and E_2 are the initial and final energies, $h = 6.55 \times 10^{-27}$ erg seconds, and ν is the frequency of the light absorbed or given out.

The atom absorbs from its environment the quanta relevant to the particular electron transfers of which it is capable at the time. These transfers are, of course, governed by the number and arrangement of the spectroscopic valency electrons, or in other words, by the state of ionization or excitation of the atom.

The unionized (or neutral) atom in the unexcited state absorbs the *ultimate lines* by the removal of one electron from its normal stationary state to some other which can be reached from that state, and re-emits them by the return of the electron to that state. The electron may, of course, leave the state to which it was carried by the ultimate absorption and pass to some state other than the normal one. If this final state is a state of higher energy than the previous state, the line produced by the process

[6] A. Fowler, Proc. Roy. Soc., 105A, 299, 1924.

12   THE LABORATORY BASIS OF ASTROPHYSICS

will be an absorption line; if it is of lower energy the result will be the production of an emission line. In either case the line produced by the transfer of an electron from a stationary state other than the normal state is known as a *subordinate line*. The

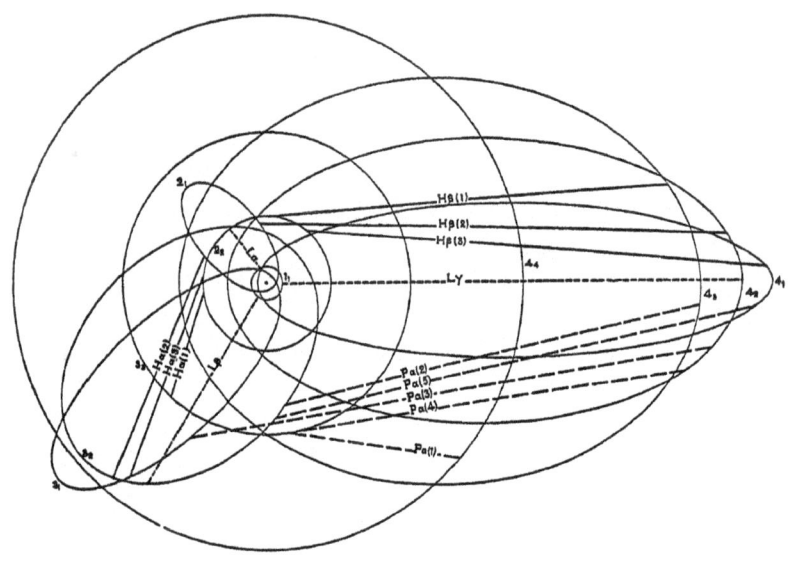

FIGURE 2

The hydrogen atom. The ten innermost orbits possible for the single electron of the atom of hydrogen are diagrammatically represented. All possible quantum transitions between the orbits are indicated as follows: — short dashes, Lyman series, terminating at a 1-quantum orbit; full lines, Balmer series, terminating at a 2-quantum orbit; long dashes, Paschen series, terminating at a 3-quantum orbit. Transfers are only possible between orbits with azimuthal quantum numbers differing by ±1.

distinction between series of ultimate and subordinate lines is of great importance in the astrophysical applications of the theory of ionization.

When the energy supply from the environment is great enough, the " outermost " (or most easily detachable) valency electron is entirely removed by the energy absorbed. In consequence the atom is superficially transformed, giving rise to a totally new spectrum, which strongly resembles the spectrum of the atom next preceding in the periodic system. Bohr's table embodies the interpretation of this resemblance — the so-called Displacement Rule of Kossell and Sommerfeld[7] — which has recently been strikingly confirmed by a very complete investigation of the arc and spark (neutral and ionized) spectra of the atoms in

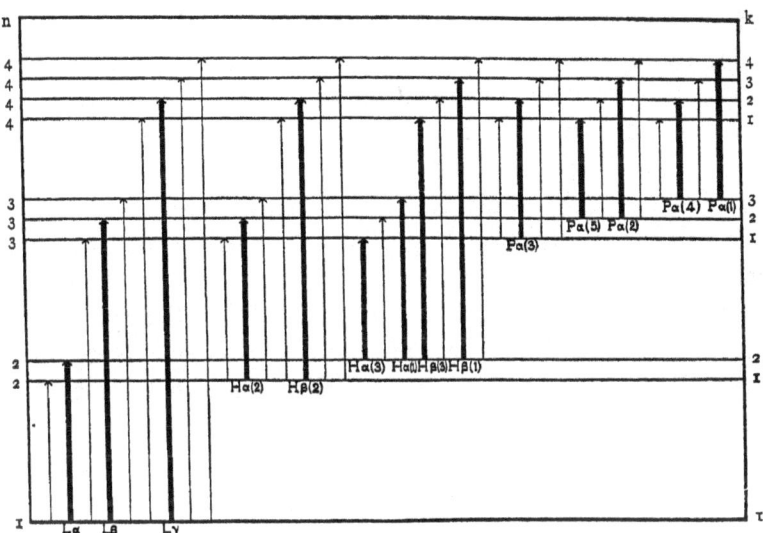

FIGURE 3

Energy levels for the hydrogen atom. Horizontal lines represent diagrammatically the levels of energy corresponding to all the possible electron orbits up to and including those of total quantum number four. Total quantum numbers are indicated on the left margin, azimuthal quantum numbers on the right margin. Transitions are only possible between orbits which differ by ± 1 in azimuthal quantum number. All such possible transitions are indicated in the diagram by heavy lines. "Forbidden jumps," for which the difference in azimuthal quantum number is zero or greater than 1, are indicated by light lines. This diagram embodies the same relations as Figure 2, the levels representing the various orbits in that figure.

the first long period.[8] It may be seen at once, for instance, that the removal of the outermost (or 3_2) electron from the atom of aluminum (13) produces an arrangement of external electrons identical with that for magnesium (12). The ionized atom produced by the complete removal of one electron gives, like the neutral atom, two kinds of line spectrum — the ultimate lines and the subordinate lines.

[7] Sommerfeld, Atombau und Spektrallinien, 3d edition, 457, 1922.
[8] Meggers, Kiess, and Walters, J. Op. Soc. Am., 9, 355, 1924.

Effectively, the ionized atom may be regarded as a new atom altogether. It reproduces the spectrum of the atom of preceding atomic number, in cases which have been fully investigated, with great fidelity, excepting that the Rydberg constant in the series formula is multiplied by four. For the twice and thrice ionized atoms the same is true, the Rydberg constant being multiplied by nine and by sixteen in the two cases. It is scarcely necessary to mention the beautiful confirmation of the theory that has been furnished by the analyses [9,10] of the spectra of Na, Mg, and Mg+, Al, Al+, and Al++, and Si, Si+, Si++, and Si+++. The attribution of the Pickering series (first observed in the spectrum of ζ Puppis) to ionized helium was the first established example of the displacement rule, and constituted one of the earliest triumphs of the Bohr theory.[11] The detection and resolution of the alternate components of that series, which fall very near to the Balmer lines of hydrogen in the spectra of the hottest stars, and the consequent derivation of the Rydberg constant for helium,[12] represents an astrophysical contribution to pure physics which is of the highest importance.

IONIZATION AND EXCITATION

The *ionization potential* of an atom is the energy in volts that is required in order to remove the outermost electron to infinity. The *excitation potential* corresponding to any particular spectral series is the energy in volts that must be imparted to the atom in the normal state in order that there may be an electron in a suitable electron orbit for the absorption or emission of that series. Several different excitation potentials are usually associated with one atom. The ionization potential and the excitation potentials are collectively termed the *critical potentials*.

From the astrophysical point of view, ionization and excita-

---

[9] A. Fowler, Report on Series in Line Spectra, 1922; Bakerian Lecture, 1924.
[10] Paschen, An. d. Phys., **71**, 151, 1923.
[11] Sommerfeld, Atombau und Spektrallinien, 3d edition, 255, 1922; A. Fowler, Proc. Roy. Soc., **90A**, 426, 1913; Paschen, An. d. Phys., **50**, 901, 1919.
[12] H. H. Plaskett, Pub. Dom. Ap. Obs., **1**, 348, 1922.

CRITICAL POTENTIALS 15

tion potentials are important as forming the basic data for the Saha theory of thermal ionization, with which the greater part of this work is concerned. A list of the ionization potentials hitherto determined is therefore reproduced in the following table. The first two columns contain the values obtained by the physical and spectroscopic methods, respectively. The third column contains "astrophysical estimates," which are inserted here to make the table more complete. The derivation of the astrophysical values will be discussed [13] in Chapter XI. Physical values result from the direct application of electrical potentials to the element in question, and spectroscopic values are derived from the values of the optical terms. (See Appendix.)

TABLE III

Atomic Number	Element	Ionization potential			Reference
		Physical	Spectroscopic	Astrophysical	
1	H	14.4, 13.3	13.54		1, 2, 3
2	He	25.4	24.47		5, 4
	He+	54.3	54.18		3, 5
3	Li		5.37		3
	Li+	40			6
4	Be		9.6		7
5	B		8.3		7
	B+		19.		7
6	C+		24.3		8
	C++			45	9, 12
7	N	16.9			10
	N+			24	9, 12
	N++			45	9, 12
8	O	15.5	13.56		3, 11
	O+			32	9, 12
	O++	50		45	12
10	Ne	16.7			13
11	Na	5.13	5.12		14, 3
	Na+	30–35			15
12	Mg	7.75	7.61		16, 3
	Mg+		14.97		3

[13] p. 156.

16 THE LABORATORY BASIS OF ASTROPHYSICS

Atomic Number	Element	Ionization potential			Reference
		Physical	Spectroscopic	Astrophysical	
13	Al		5.96		3
	Al+		18.18		17
	Al++		28.32		17
14	Si		10.6	8.5	18, 19
	Si+		16.27		18
	Si++		31.66		18
	Si+++		44.95		18
15	P	13.3, 10.3			20, 21
	P++		29.8		7
	P+++		45.3		7
16	S	12.2	10.31		20, 11
	S+			20	9, 12
	S++			32	9, 12
	S+++		46.8		7
17	Cl	8.2			13
18	A	15.1			22
	A+	33, 34, 41.5			23, 22, 24
19	K	4.1	4.32		14, 3
	K+	20–23			15
20	Ca		6.09		3
	Ca+		11.82		3
21	Sc			6–9	25
	Sc+			12.5	19
22	Ti		6.5		26
	Ti+			12.5	19
23	V			6–9	25
24	Cr		6.7		3
25	Mn		7.41		35
26	Fe		5.9, 8.15	7.5	28, 29, 19
	Fe			13	19
27	Co			6–9	25
28	Ni			6–9	25
29	Cu		7.69		3
30	Zn		9.35		3
	Zn+		19.59		7
31	Ga		5.97		3
33	As	11.5			30
34	Se	12–13, 11.7			31, 32
35	Br	1.00			13
36	Kr	14.5			33
37	Rb	4.1	4.16		34, 3

IONIZATION POTENTIALS OF ATOMS

Atomic Number	Element	Ionization potential			Reference
		Physical	Spectroscopic	Astrophysical	
38	Sr		5.67		3
	Sr+		10.98		3
42	Mo		7.1, 7.35		35, 36
47	Ag		7.54		3
48	Cd		8.95		3
	Cd+		18.48		7
49	In		5.75		37
51	Sb	8.5 ± 1.0			26
53	I	10.1, 8.0			38, 39
56	Ba		5.19		3
	Ba+		9.96		3
80	Hg		10.4		40
81	Tl		6.94		41
82	Pb	7.93	7.38		42
83	Bi	8.0			30
	Bi+	14.			30

[1] Horton and Davies, Proc. Roy. Soc., **97A**, 1, 1920.
[2] Mohler and Foote, J. Op. Soc. Am., **4**, 49, 1920.
[3] A. Fowler, Report on Series in Line Spectra, 1922.
[4] Lyman, Phys. Rev., **21**, 202, 1923.
[5] Horton and Davies, Proc. Roy. Soc., **95A**, 408, 1919.
[6] Mohler, Science, **58**, 468, 1923.
[7] D. R. Hartree, unpub.
[8] A. Fowler, Proc. Roy. Soc., **105A**, 299, 1924.
[9] Payne, H. C. 256, 1924.
[10] Brandt, Zeit. f. Phys., **8**, 32, 1921.
[11] Hopfield, Nature, **112**, 437, 1923.
[12] R. H. Fowler and Milne, M. N. R. A. S., **84**, 499, 1924.
[13] Horton and Davies, Proc. Roy. Soc., **98A**, 121, 1920.
[14] Tate and Foote, Phil. Mag., **36**, 64, 1918.
[15] Foote, Meggers, and Mohler, Ap. J., **55**, 145, 1922.
[16] Foote and Mohler, Phil. Mag., **37**, 33, 1919.
[17] Paschen, An. d. Phys., **71**, 151 and 537, 1923.
[18] A. Fowler, Bakerian Lecture, 1924.
[19] Menzel, H. C. 258, 1924.
[20] Mohler and Foote, Phys. Rev., **15**, 321, 1920.
[21] Duffendack and Huthsteiner, Amer. Phys. Soc., 1924.
[22] Horton and Davies, Proc. Roy. Soc., **102A**, 131, 1922.
[23] Shaver, Trans. Roy. Soc. Can., **16**, 135, 1922.
[24] Smyth and Compton, Amer. Phys. Soc., 1925.
[25] Russell, Ap. J., **55**, 119, 1922.
[26] Kiess and Kiess, J. Op. Soc. Am., **8**, 609, 1924.
[27] Catalan, Phil. Trans., **223A**, 1922.
[28] Sommerfeld, Physica, **4**, 115, 1924.
[29] Giesler and Grotrian, Zeit. f. Phys., **25**, 165, 1924.
[30] Ruark, Mohler, Foote, and Chenault, Nature, **112**, 831, 1923.
[31] Foote and Mohler, The Origin of Spectra, 67, 1922.
[32] Udden, Phys. Rev., **18**, 385, 1921.
[33] Sponer, Zeit. f. Phys., **18**, 249, 1923.
[34] Foote, Rognley and Mohler, Phys. Rev., **13**, 61, 1919.
[35] Catalan, C. R., **176**, 1063, 1923.
[36] Kiess, Bur. Stan. Sci. Pap. 474, 113, 1923.
[37] McLennan, Br. A. Rep., **25**, 1923.
[38] Foote and Mohler, The Origin of Spectra, 67, 1922.
[39] Smyth and Compton, Phys. Rev., **16**, 502, 1920.
[40] Eldridge, Phys. Rev., **20**, 456, 1922.
[41] Mohler and Ruark, J. Op. Soc. Am., **7**, 819, 1923.
[42] Grotrian, Zeit. f. Phys., **18**, 169, 1923.

By the use of one or other of the available methods, the data for neutral atoms are complete as far as atomic number 38, with the exception of carbon (6), fluorine (9) and germanium (32). The data for ionized atoms are also increasing, at the present time, in a very gratifying manner. The " hot spark " investigations of Millikan and Bowen,[14] which permit the estimation of the fifth and sixth ionization potentials of certain light atoms, are not included in the table. Under the conditions hitherto investigated in the stellar atmosphere, ionization corresponding to a potential of about fifty volts is the highest encountered, and accordingly ionization potentials that greatly exceed this value have no place in the present tabulation of astrophysically useful data. A knowledge of the higher critical potentials [15] is, however, of great interest in connection with the theoretical problems of the far interior of the star.

There are conspicuous gaps in the table, and it is to be feared that many of them are likely to remain unfilled. The spectra of the neutral atoms of carbon, phosphorus, and nitrogen have hitherto defied analysis, and our knowledge of the corresponding ionization potentials must therefore depend on physical methods. For carbon, silicon, and similar refractory materials, such methods are difficult of application; the same applies to the metals. It is therefore probable that the ionization potentials of the neutral atoms of several of the lighter elements, of the platinum metals, and of the rare earths, will remain unknown or uncertain for some time to come. None of the atoms thus omitted is of immediate astrophysical importance.

As shown in the table, the values for the ionized and doubly ionized light atoms O+, O++, C++, N++, S+, and S++ are deduced only astrophysically. It may be hoped that the spectra of these atoms will soon be arranged in series, so that an accurate value of the ionization potential may be available, in place of the approximate one deduced from the stellar evidence,

---

[14] Millikan and Bowen, Phys. Rev., 23, 1, 1924; Nature, 114, 380, 1924.
[15] Hartree, Proc. Camb. Phil. Soc., 22, 464, 1924; Thomas, Phys. Rev., 25, 322, 1925.

for the corresponding absorption lines are of importance in the spectra of the hotter stars.

The spectroscopic ionization potentials have an advantage over the physical values, in that the corresponding state of the atom is known with certainty, whereas physical methods can in general only *detect* some critical potential, without assigning it definitely to a particular transition. For example, it seems likely that in some cases the first ionization, whether caused by incident radiation or by electron impacts, corresponds to the loss of an electron by the *molecule*:

$$E_2 = E_2^+ + e$$

where E represents the atom, and e the electron. The effect of increased excitation would then be the decomposition

$$E_2^+ = E^+ + E$$

The first reaction would produce the ionized molecule, and the second would produce the ionized and neutral atoms *simultaneously*. It might thus happen that the E^+ spectrum could appear without the previous appearance of the E spectrum, since all of the element was present in the form E_2 before ionization.

The above is only a simple illustrative example of the possible complexity in the physical determination of ionization potentials. The interpretation of four successive critical potentials for hydrogen has been discussed by Franck, Knipping and Krüger,[16] while eight have been detected by Horton and Davies[17] for the same element. Similarly Smyth[18] discusses four critical voltages for nitrogen. No explicit attempt has yet been made to use these facts for the interpretation of astrophysical data, but they may account for the unexplained absence of some neutral elements from the cooler stars. The absence is generally to be attributed, as will be shown in Chapter V, to the non-occurrence of suitable lines in the part of the spectrum usually ex-

---

[16] Verh. d. Deutsch. Phys. Ges., 21, 728, 1919.
[17] Phil. Mag., 46, 872, 1923.
[18] Proc. Roy. Soc., 103A, 121, 1923.

amined. But it is possible that the persistence of the molecule has a definite significance in the case of nitrogen, where the ionization potential is as high as 16.9 volts.

The increasing completeness of the table of ionization potentials suggests a re-examination of the relation recently traced by

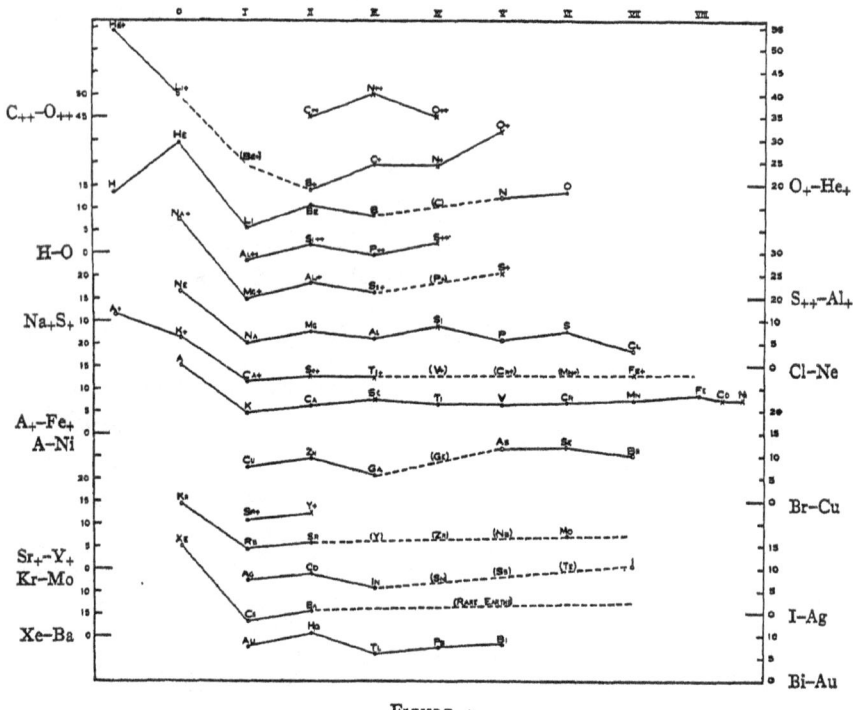

FIGURE 4

Relation between ionization potential and position in the periodic system. Ordinates are ionization potentials in volts, on the equal but shifted scales indicated alternately on left and right margins. Abscissae are columns of the periodic table. Physical determinations of ionization potential are indicated by open circles; dots give spectroscopic determinations, and crosses denote astrophysical estimates. Conjectural portions of the curve are indicated by broken lines, and atoms of unknown ionization potential are enclosed in parentheses.

the writer [19] between ionization potential and atomic number. The original diagram, in which columns of the periodic table are treated as abscissae, and the ordinates are ionization potentials on equal but shifted scales, so that analogous elements fall one

[19] Proc. N. Ac. Sci., 10, 322, 1924.

below another, is here reproduced, with the addition of data more recently obtained.

The Displacement Rule of Kossell and Sommerfeld leads us to expect a pronounced similarity between the line drawn in the diagram from the point representing one element to that representing the next, and the corresponding line for the ionized atoms of the same elements, the latter being shifted one place to the left for each electron removed. The points for once and twice ionized atoms are inserted into the diagram on this principle, and the parallelism is found to exist. The regularities of the diagram and their possible significance (such, for example, as the pairing of the valency electrons, the second being harder to remove than the first) were discussed in the original paper. All the more recent data appear to confirm the conclusion there set forth, that the relation between ionization potential and atomic number is very closely the same in each period.

Duration of Atomic States

In addition to the critical potentials, which give a measure of the ease with which an atom is excited or ionized, astrophysical theory requires an estimate of the readiness with which an atom recovers after excitation or ionization. It appears probable that this factor, like the critical potentials, is independent of external conditions, and depends upon something that is intrinsic in the atomic structure. The "life" of the atom has been extensively investigated in the laboratory, and has been shown to be a small fraction of a second in duration. Probably this subject of "atomic lives" is still in an initial stage, and the accuracy of the results and the range of elements discussed will be greatly increased in the near future. A summary of the material obtained up to the present time is contained in the following table. Successive columns contain the atom discussed, the deduced atomic life in seconds, the authority, and the reference.

The data are practically confined to hydrogen and mercury, and for both these elements the atomic life appears to be of the order 10^{-8} seconds.

Astrophysical estimates of the life of the excited calcium atom have been made by Milne,[20] who derives values of the order 10^{-8} seconds. This is so near to the values obtained in the laboratory that it seems permissible, in the absence of further precise data,

TABLE IV

Atom	Life	Authority	Reference
Hα	1.5×10^{-8}	Wien	An. d. Phys., 60, 597, 1919
Hβ	1.6×10^{-8}	Ibid.	Ibid.
Hγ	1.6×10^{-8}	Ibid.	Ibid.
O+	1.5×10^{-8}	Ibid.	Ibid.
Hβ	5×10^{-8}	Dempster	Phys. Rev., 15, 138, 1920
Hg	7×10^{-5}	Wood	Proc. Roy. Soc., 99 A, 362, 1921
Hg	10^{-8}	Franck and Grotrian	Zeit. f. Phys., 4, 89, 1921
Hβ, Hγ	2.3×10^{-8}	Mie	An. d. Phys., 66, 237, 1921
Hβ, Hγ	2.3×10^{-8}	Wien	An. d. Phys., 66, 232, 1921
N bands	3.1×10^{-8}	Ibid.	Ibid.
Hα, Hβ	1.8×10^{-8}	Ibid.	An. d. Phys., 73, 483, 1924
He 4478	1.8×10^{-8}	Ibid.	Ibid.
Hg 4358	1.8×10^{-8}	Ibid.	Ibid.
Hg 2536	9.7×10^{-8}	Ibid.	Ibid.
Hg	10^{-7}	Turner	Phys. Rev., 23, 464, 1924
Hg	10^{-6}	Webb	Phys. Rev., 21, 479, 1923

to assume an atomic life of 10^{-8} seconds, as a working hypothesis, for all atoms. The same value is unlikely to obtain for all atoms; in particular it may be expected to differ for atoms in different states of ionization. But here astrophysics must be entirely dependent on further laboratory work for the determination of a quantity that is of fundamental importance.

RELATIVE PROBABILITIES OF ATOMIC STATES

The relative intensities of lines in a spectrum must depend fundamentally upon the relative tendencies of the atom to be in the corresponding states. To a subject which, like astrophysics,

[20] Proc. Phys. Soc. Lond., 36, 94, 1924.

PROBABILITIES OF ATOMIC STATES   23

depends for its data largely upon the relative intensities of spectral lines, the theory of the relative probabilities of atomic states is of extreme importance. The question is obviously destined to become an important branch of spectrum theory. It has been discussed, from various aspects, by Füchtbauer and Hoffmann,[21] Einstein,[22] Füchtbauer,[23] Kramers,[24] Coster,[25] Fermi,[26] and Sommerfeld.[27] The comparison with observation has been made, up to the present, only for a few elements. The relative intensities of the fine-structure components of the Balmer series of hydrogen were examined by Sommerfeld,[28] and exhaustive work with the calcium spectrum has recently been carried out by Dorgelo.[29] The astrophysical application of the data bearing on relative intensities of lines in the spectrum of one and the same atom, while an essential branch of the subject, is a refinement which belongs to the future rather than to the present.

EFFECT ON THE SPECTRUM OF CONDITIONS AT THE SOURCE

(a) *Temperature Class.* — It is found experimentally that the relative intensities of the lines in the spectrum of a substance are altered when the temperature is changed. Some lines, notably the ultimate lines mentioned in a previous paragraph, predominate at low temperature. Other lines, which are weak under these conditions, become stronger if the temperature is raised, and lines which are the characteristic feature of the spectrum at the highest temperatures that can be attained in the furnace are often imperceptible at the outset. The effects are more conspicuous, and have been most widely studied, in the spectra of the metals, which are rich in lines and are amenable

[21] An. d. Phys., **43**, 96, 1914.
[22] Phys. Zeit., **18**, 121, 1917.
[23] Phys. Zeit., **21**, 322, 1922.
[24] Proc. Copenhagen Ac., 1919.
[25] Physica, **4**, 337, 1924.
[26] Physica, **4**, 340, 1924.
[27] Zeit. f. Tech. Phys., **5**, 2, 1925.
[28] Atombau und Spektrallinien, 3d edition, 588, 1922.
[29] Physica, **3**, 188, 1923; Zeit. f. Phys., **13**, 206, 1923; *ibid.*, **22**, 270, 1924; Dissertation, Utrecht, 1924; Physica, **5**, 27, 1925.

to furnace conditions. The results of such experiments, which are chiefly the work of A. S. King, are expressed by the assignment of a "temperature class," ranging from I to V, to each line; Class I represents the lines characteristic of the lowest temperatures, and Class V denotes the lines that require the greatest stimulation.

The temperature class of a line is intimately connected with the amount of energy required to excite the line. It may, indeed, be used as a rough criterion of excitation potential, high temperature class indicating high excitation energy. The temperature class is therefore useful in assigning series relations to unclassified lines, and is of value to the astrophysicist chiefly in this capacity of a classification criterion. King's work on silicon shows, for instance, that 3906 is of Class II, and is therefore not an ultimate line — a fact which has considerable significance in studying the astrophysical behavior of the line.

The correlation of temperature class with excitation potential receives an immediate explanation in terms of the theory of thermal ionization. It furnishes a useful laboratory corroboration of the theory by showing that the thermal excitation of successive lines, with rising excitation potential, takes place in qualitative agreement with prediction.

The appended list shows the atoms for which the spectra have been analyzed by King on the basis of temperature class:

Element	Reference	Element	Reference
Iron	Mt. W. Contr. 66, 1912	Calcium	Mt. W. Contr. 150, 1918
Titanium	Mt. W. Contr. 76, 1914	Strontium	Ibid.
Vanadium	Mt. W. Contr. 94, 1914	Barium	Ibid.
Chromium	Ibid.	Magnesium	Ibid.
Cobalt	Mt. W. Contr. 108, 1915	Manganese	Mt. W. Contr. 198, 1920
Nickel	Ibid.	Silicon	Pub. A. S. P., 22, 106, 1921

(b) *Pressure.* — In the laboratory the observed effects of pressure [30] are a widening and shifting of the lines in the spectrum — effects which differ in magnitude and direction for dif-

[30] King, Mt. W. Contr. 53, 1911; *ibid.*, 60, 1912.

SPECTRAL EFFECTS OF PHYSICAL FACTORS 25

ferent lines. The phenomena are well marked under pressures of several atmospheres.

Recent developments of astrophysics, such as are summarized in Chapter III and Chapter IX, have shown that the pressures in stellar atmospheres are normally of the order of a hundred dynes per square centimeter, or less. At such pressures no appreciable pressure shifts will occur, and indeed one of the most direct methods by which these exceedingly low pressures in reversing layers have been established [31] is based on the absence of appreciable pressure effects.

(c) *Zeemann Effect.* — The magnetic resolution of spectral lines into polarized components [32] has, as yet, for the astrophysicist, chiefly a value as a criterion for classifying spectra. In the field of solar physics proper, a direct study of the Zeemann effect has led to important results.[33] The present study is not, however, explicitly concerned with the sun, except in comparing solar features with similar features that can also be examined in the stars.

The investigations of Landé on term structure and Zeemann effect [34] for multiplets have shown how the Zeemann pattern formed by the components into which a line is magnetically resolved can be related to the series attribution of the line. This provides a method of classifying spectra which are rich in multiplets, and which have previously defied analysis. The indirect astrophysical value of the Zeemann effect is, therefore, very great.

(d) *Stark Effect.* — The effect of an electric field in resolving spectral lines into polarized components was first pointed out by Stark [35] for hydrogen and helium. Several other investigators have since studied the effect for these two elements,[36] and for

[31] St. John and Babcock, Ap. J., 60, 32, 1924.
[32] Zeemann, Researches in Magneto-Optics, 1911.
[33] Hale, Mt. W. Contr. 30, 1908.
[34] Landé, Zeit. f. Phys., 15, 189, 1923.
[35] Stark, Elektrische Spektralanalyse Chemischer Atome, 1914.
[36] Merton, Proc. Roy. Soc., 92A, 322, 1915; *ibid.*, 95A, 33, 1919.

various metals.[37,38] Unlike the temperature and magnetic effects, the Stark effect has not been used as a criterion for the series relations of unclassified lines.

The Stark effect has not been detected in the solar spectrum, presumably because the concentration of free electrons prevents the formation of large electrostatic fields.

Several investigators, however, have contemplated in the Stark effect a possible factor influencing the stellar spectrum.[39,40] It does not seem unlikely that nuclear fields could operate as a sensible general electrostatic field at the photospheric level, thus producing a widening and winging of certain lines. The question has been numerically discussed by Hulburt,[41] and Russell and Stewart,[42] in an examination of Hulburt's work, concluded that the Stark effect might possibly make some contribution (probably not a preponderant one) to the observed widths of lines in the solar spectrum. The question is not definitely settled, but it appears well to keep so important a possibility in mind.

[37] Anderson, Mt. W. Contr. 134, 1917.
[38] Takamine, Mt. W. Contr. 169, 1919.
[39] Evershed, Observatory, 45, 166, 1922; *ibid.*, 45, 296, 1922.
[40] Lindemann, Observatory, 45, 167, 1922.
[41] Hulburt, Ap. J., 59, 177, 1924.
[42] Russell and Stewart, Ap. J., 59, 197, 1924.

CHAPTER II

THE STELLAR TEMPERATURE SCALE

IT is well to distinguish the different meanings that are to be associated with the term "stellar temperature." The observed energy distribution in the spectrum, combined with the theory of black-body radiation, lead to a quantity known as the "effective temperature" of the star. This is the temperature of a hypothetical black body, the spectrum of which would have the observed energy distribution of the star in question. It has often been emphasized that the effective temperature is merely a label, for it is not the actual temperature of any specific portion of the star. Presumably the temperature of a star falls off, from the center outwards, according to the laws expressed by the theory of radiative equilibrium, and though it might thus be possible to specify, on certain assumptions, the depth in a star at which the effective temperature coincides with the actual temperature, no observational significance could attach to the information.

The theory of radiative equilibrium [1] enables us to specify the temperature gradient, and in particular to determine the central temperature, the effective temperature, and the boundary temperature, corresponding to a given energy output. These three quantities are essentially arbitrary, and the second is the only one susceptible of direct measurement, while none of them represents the actual temperature of any assignable region. In order to clarify ideas it is useful to regard the effective temperature as representing roughly the temperature of the photosphere, that is, of the region in the star that gives rise to the approximately black continuous background of the spectrum. It must, however, be remembered that "the theory provides a definite relation between temperature and optical depth, involving only

[1] Eddington, Zeit. f. Phys., **7**, 351, 1921.

one constant, the effective temperature. Suppose now . . . we arbitrarily select a certain temperature, and name it the photospheric temperature, and name the unknown depth at which it occurs the photospheric depth; this depth will be described by some unknown transmission coefficient, to be determined. If, taking account of absorption and emission, we proceed to calculate the transmission coefficient . . . we shall simply recover the optical depth predicted by Schwarzschild's theory." (Milne.)[2] No method of measuring the effective temperatures of the stars by comparing their energy spectrum with that of a black body can remove the arbitrariness of the quantity thus measured.

The theory of thermal ionization permits estimates to be made of the temperatures in the reversing layers of stars. These temperatures refer to the average level at which are situated the absorbing atoms corresponding to the lines used. The differences of effective level[3] for different atoms render these "ionization temperatures" difficult to define consistently, but they represent actual temperatures of assignable regions in the star, and the extent of their agreement with the temperatures derived from the distribution of energy in the continuous spectrum is a matter of extreme interest. The material and theory from which the ionization temperatures are derived is the subject matter of Chapters VI to IX. The temperature scale used in calibration and in the discussion of the theory of thermal ionization is the scale derived from the measured *effective temperatures*.

The derivation of a definitive scale of effective temperatures from the numerous available observations is probably impossible at the present time. The methods employed differ widely, and the conditions for accurate intercomparison cannot be regarded as fully established. The material at present available, however, permits some general conclusions, and as the needs of astrophysics demand a *working* temperature scale, such conclusions are summarized in the present chapter.

[2] Phil. Trans., 223 A, 201, 1922.
[3] Chapter IX, p. 136.

COMPARISON OF TEMPERATURE SCALES 29

In the discussion of the material a difficulty immediately arises. The scale to be derived must be based entirely, in the present stage of the observations, upon the apparently brighter stars, and it is notorious that they are not homogeneous in absolute magnitude. Theory predicts [4] that absolutely bright stars will have a lower effective temperature than stars of low luminosity belonging to the same spectral class, and this prediction is, on the whole, verified by observation. The material must therefore be selected on the basis of luminosity if a standard temperature scale is to be formed, and probably the temperature scale to be aimed at should refer to stars of some one absolute magnitude adopted as standard. Theoretically, standard mass might be preferable to standard luminosity, but, in the present state of the subject, so few masses are known that such a system would not be practicable. The ideal of referring to standard absolute magnitude was not attained by the earlier temperature scales, which were apparently based upon averages for all the available brighter stars.

The more comprehensive data for the study of the stellar temperature scale are the spectrophotometric measures of Wilsing and Scheiner,[5] of Wilsing,[6] of E. S. King,[7] and of Rosenberg.[8] The temperature scales derived by Wilsing and by Rosenberg differ by a linear factor; Rosenberg assigns higher temperatures to the hotter stars, and lower temperatures to the cooler stars. These temperature scales, and their intercomparison, have been very fully discussed by Brill,[9] who reduces all the measures to the scale given by Wilsing, and gives, for the principal Draper classes, the following comparative table for the corrected mean effective temperatures on the absolute centigrade scale.

In addition to the comprehensive data just quoted, there

[4] Chapter XIV, p. 195.
[5] Wilsing and Scheiner, Pots. Pub., 24, No. 74, 1919.
[6] Pots. Pub., 24, No. 76, 1920.
[7] H. A., 76, 107, 1916.
[8] A. N., 193, 356, 1912.
[9] A. N., 218, 210, 1923; *ibid.*, 219, 22 and 354, 1923; Die Strahlung der Sterne, Berlin, 1924.

have been numerous determinations of the temperatures of individual bright stars, chiefly by Abbot,[10] Coblentz,[11] Sampson,[12] and H. H. Plaskett.[13] In the main these values confirm the scale given in Table V, but sometimes considerable differences

TABLE V

Class	Wilsing	Rosenberg	E. S. King Color Temperature	E. S. King Total Radiation
B0	12300°	30000°	22700°	22700°
B5	11450	18000	15200	14900
A0	10250	12000	11600	11300
A5	9000	9000	8800	8600
F0	7950	7850	7900	7700
F5	6880	6930	7000	6800
G0	5980	6000	6040	5870
G5	5250	5200	5090	4950
K0	4570	4570	4570	4440
K5	3860	3840	3640	3550
Ma	3550	3580	3430	3340

occur in the values given for individual stars by different investigators. At the same time, each observer is usually reasonably self-consistent, and the deviations must therefore be ascribed to differences of method. Some of the results are reproduced, for illustration, in Table VI.

It is seen that the effective temperatures of individual hotter stars vary widely among themselves. This is largely a result of the difficulty of making the appropriate correction for atmospheric extinction. It must, then, be supposed that the temperatures derived by spectrophotometric methods are not trustworthy for stars hotter than Class A5. The values determined by the earlier observers for the A and B classes are almost certainly too low. Rosenberg's value of 30,000° for B0 is, however, most probably too high, as will be inferred later from the ionization temperature scale.

For the cooler stars small discrepancies also occur among the different observers. In the writer's opinion, the lowest estimates

[10] Rep., Smithsonian Ap. Obs., 1924.
[11] Pop. Ast., 21, 105, 1923.
[12] M. N. R. A. S., 85, 212, 1925.
[13] Pub. Dom. Ap. Obs., 2, 12, 1923.

for the temperatures of the cooler stars are probably nearest to the truth.

TABLE VI

Star	Abbot Radiometric	Coblentz Thermoelectric	Plaskett Wedge Method	Sampson Photoelectric
ε Ori(B0)		13000°		25000°
γ Cas(B0p)			15000°	30000
ε Per(B0)			15000	14000
β Ori(B8)	16000°	10000		14800
α Lyr(A0)	14000	8000		11600
α CMa(A0)	11000			12800
α Cyg(A2p)		9000	9000	10900
α Aql(A5)		8000		
δ Cas(A5)			9000	10700
α CMi(F5)		6000		8300
α Aur(G0)	5800	6000	5500–6000	5500*
α Boo(K0)		4000		4200
β Gem(K0)		5500	5000–5500	4200
α Tau(K5)	3000	3500		3400
α Ori(Ma)	2600	3000		3400
β Peg(Mb)	2850			3200

* Temperature assumed in calibration of scale.

It was mentioned at the outset that dwarf stars appear to be at a higher temperature than giants of the same spectral class. The following table summarizes the differences in temperature, as compiled by Seares.[14]

TABLE VII

Class	Effective Temperature	
	Giant	Dwarf
F5	6080°	6080°
G0	5300	5770
G5	4610	5500
K0	3860	4880
K5	3270	4120
Ma	3080	3330

A more detailed list of giant and dwarf temperatures was compiled in 1922 by Hertzsprung[15] from all the material then available. The tabulation that follows contains his values for

[14] Ap. J., 55, 202, 1922.   [15] Lei. An., 14, 1, 1922.

c_2/T (the "reciprocal temperature," where c_2 is 14.6??, and ? corresponding absolute temperature, in degrees centigrade

TABLE VIII

Mt. W. Class	c_2/T Giant	c_2/T Dwarf	Temperature Giant	Temperature Dwarf
A5		2.05		7300
A6		2.10		6900
A7		2.05		6900
A8		2.10		6900
A9		2.30		7400
F0		2.33		
F1		2.30		
F2		2.30		
F3		2.33		
F4		2.30		
F5		2.45		
F6	2.30	2.53	6340	
F7		2.45		
F8		2.71		
F9	2.83	2.60	5150	
G0	2.92	2.68	5000	5640
G1	2.92	2.64	5020	
G2	3.13		4710	
G3	3.00		4850	
G4	3.15		4700	
G5	3.25	2.76	4480	5300
G6	3.20		4540	
G7	3.20		4480	
G8	3.30	3.04	4030	4850
G9	3.48	3.13	4180	4700
K0	3.50	3.05	4170	4850
K1	3.54		4130	
K2	3.83		3840	
K3	3.86		3790	
K4	4.14		3530	
K5	4.33		3370	
K6	4.30		3400	
K8	4.45		3400	
Ma	4.49		3250	
Mb	4.45		3280	
Mc	3.93		3710	

IONIZATION TEMPERATURE SCALE   33

The difference in temperature between giant and dwarf stars of the same spectral class is clearly shown in the foregoing tables. The relation of absolute magnitude to effective temperature within a given class must be regarded as definitely established by observation.

The temperatures for the cooler giant stars in both these lists are somewhat lower than those given for the corresponding classes in Table V. The temperature of K0, for instance, is

TABLE IX

Class	Temperature	Class	Temperature
Ma	3000°	A3	9000°
K5	3000	A0	10000
K2	3500	B8	13500
K0	4000	B5	15000
G5	5000	B3	17000
G0	5600	B1.5	18000
F5	7000	B0	20000
F0	7500	O	25000
A5	8400		35000

placed nearer to 4000° than to 4500°. The fact that the sun, a typical G0 dwarf, has an effective temperature of 5600° seems to favor these lower values.

In concluding the summary of stellar temperatures, the ionization temperature scale is given in the foregoing table. The discussion on which the table is based is contained in Chapters VI to IX, and it is merely placed here for comparison with the preceding tabulations.

CHAPTER III

PRESSURES IN STELLAR ATMOSPHERES

THE theory of thermal ionization enables us to make an analysis of the spectrum of the stellar reversing layer by predicting the number of atoms of any given kind that will be effective in absorbing light from the interior of the star, under given conditions, and by comparing the predicted values with the observed intensities of the corresponding absorption lines. The results depend partly on definite physical constants associated with the atoms — the ionization and excitation potentials, and the arrangement of the electrons around the nucleus. The temperature and pressure of the region in which the atom is situated are also required before the theory can be applied. The scale of stellar temperatures was discussed in the preceding chapter and the present chapter is devoted to a synopsis of the modern views as to pressures in the reversing layer.

Strictly speaking, we cannot refer to " the pressure in the reversing layer," for, like the temperature, the pressure has a gradient throughout the star. This gradient, as derived from the theory of radiative equilibrium,[1] is steep in the far interior of the star, but towards the outside the rapid fall of pressure begins to decrease, and changes somewhat abruptly to a very small gradient in the photospheric region, where radiation pressure and gravitation are of the same order of magnitude. Outside this layer of transition between the region dominated by radiation pressure and the region dominated by gravitation, the pressure gradient is very shallow, and decreases until, in the tenuous outer regions of the star, there is no appreciable pressure gradient, and atoms are practically floating freely.

[1] Eddington, Zeit. f. Phys., 7, 731, 1921.

The outermost regions of the atmosphere, at these exceedingly low pressures, make little or no contribution to the ordinary stellar spectrum; they can only be studied in the high-level chromosphere by means of the flash spectrum obtained at a total eclipse of the sun. The spectra that are ordinarily examined are from a region that is at an appreciable depth within the star — the depth from which the light of each individual wave-length can penetrate. The "layer" of which we can obtain a spectrum is therefore not at the same depth for all frequencies; it is most deep-seated in regions of continuous background, and nearest to the surface of the star at the centers of strong absorption lines. The pressures from which the different parts of the spectrum originate differ in the same way, and the idea of "pressure in the reversing layer" is not an easy one to define significantly.

For theoretical purposes it is usual to deal with the pressure at a given "optical depth" (a measure of the *amount of absorbing matter* traversed by the radiation in coming from the level considered). The optical depth τ is connected with the density ρ, the mass coefficient of absorption for unit density, k, and the vertical depth, z, in the star, by the relation [2]

$$k\rho\, dz = d\tau$$

The density gradient is thus eliminated. The optical depth is furthermore related to the pressure p by the relations

$$dp/dz = g\rho$$

where g is the value of gravity at the point in question, and

$$dp/d\tau = g/k$$

whence:

$$p = (g/k)\tau$$

In considering the stellar atmosphere we are dealing with a layer so near the surface that the value of g involved is effectively the "surface gravity" for the star. If k is constant, a

[2] Pannekoek, B. A. N., 19, 1922.

condition probably approximately fulfilled,[3] the pressure at constant optical depth is then directly proportional to the surface gravity, which varies as the product of the mean density and the radius of the star. Some idea of the *range* in pressure with which we shall be concerned in the stellar atmosphere can therefore be obtained from stars of known mean density and radius.

The data for eight such stars, all of the second type, are contained in Table X, which is adapted from tabulations given by Shapley.[4] Successive columns contain the name of the star, the

TABLE X

Star	Class	Mean Density		Radius		Product	
SX Cas	G3	0.0004	0.0002	15.3	18.6	0.006	0.004
RX Cas	K0	0.0005	0.0004	14.3	14.3	0.007	0.006
RZ Oph	F8	0.001	0.00003	10.1	33.5	0.010	0.001
RT Lac	G5	0.013	0.010	4.6	4.6	0.059	0.046
W Cru	Gp	0.000002	0.000025	94	36	0.00019	0.0009
U Peg	F3	0.83	0.67	1.2	1.2	1.0	0.8
W U Ma	G	1.8	1.8	0.9	0.9	1.6	1.6
Sun	G0	1.0		1.0		1.0	

spectral class, the mean densities of the two components in terms of the solar density, the hypothetical radii of the two components (on the assumption of solar mass) in terms of the sun's radius, and the product of mean density and radius for each component.

In mean density these stars display a range of 10^6, while the range in surface gravity is 10^4, illustrating the significant fact that the mean density varies far more widely than the surface gravity. The latter quantity is the important one in determining the pressure that may be assumed to exist in the reversing layer. If the masses of the very luminous stars of low mean density, such as W Crucis, exceed the solar mass, as they most

---

[3] Milne, Phil. Mag., **47**, 217, 1924.
[4] Ap. J., **42**, 271, 1915; Princeton Contr. No. 3, 82, 1915.

METHODS OF ESTIMATING PRESSURE

probably do, the hypothetical radii are increased, and the range in surface gravity becomes even smaller than before.

The data for stars of known mean density and radius permit the estimation of the range in surface gravity, and hence of the *range* in pressure, encountered in the reversing layer. In the absence of knowledge of the appropriate optical depth, however, the *actual* pressure cannot be deduced from such considerations, and recourse must be made to more indirect methods. The present view is based upon a number of considerations, none of which would alone be of great weight. All of the conclusions, taken together, however, indicate that the upper limit of the pressure for the region in which the Fraunhofer lines originate is of the order of 10^{-4} atmospheres.

Attention was first called to the probability of an extremely low pressure in the reversing layer by R. H. Fowler and Milne,[5] in advancing the form of ionization theory which is to be analyzed in later chapters. The conclusion that the pressure in the reversing layer is exceedingly low was a direct outcome of their discussion, and they mentioned that the results from other methods converged in the same direction.

Russell and Stewart,[6] in a specific discussion of the pressures at the surface of the sun, have established beyond question, and on quite other grounds, that the pressure for the solar reversing layer is indeed of the order suggested by Fowler and Milne. The value 10^{-4} atmospheres need then no longer be regarded as a *result* of the Fowler-Milne theory, and may be used without redundancy in deriving a stellar temperature scale from that theory.

METHODS OF ESTIMATING REVERSING LAYER PRESSURES

Russell and Stewart examined the evidence for reversing layer pressures derived from the following sources: (*a*) Shifts of spectral lines due to pressure, (*b*) Sharpness of lines, (*c*) Widths of lines, (*d*) Flash spectrum, (*e*) Equilibrium of outer layers, (*g*)

[5] M. N. R. A. S., **83**, 403, 1923.   [6] Ap. J., **59**, 197, 1924.

Ionization and chemical equilibrium in the solar atmosphere. In addition to these we have (f) the observed limit of the Balmer series in the hotter stars, where the hydrogen lines are at or near their maximum. These sources of evidence will now be briefly discussed.

(a) *Shifts of Spectral Lines.* — It was at one time supposed that displacements of spectral lines, corresponding to pressures of several atmospheres, could be found in stellar spectra. More recent work,[7] however, has shown conclusively that the pressure shifts that occur are so small that it is impossible to estimate a pressure from them with any approach to accuracy. The estimated pressures are of the same order as their probable errors. This being so, the most that can be expected of the method based upon pressure effects is a demonstration of whether or no the pressure exceeds 0.1 atmosphere, and this question has now been satisfactorily answered in the negative.

(b) *Sharpness of lines.* — The occurrence, as sharp distinct lines in the spectra of the stellar atmosphere, of lines that are diffuse in the laboratory at atmospheric pressure, and only become sharp when the pressure is very much reduced, indicates that the pressure in the reversing layer must be extremely low. The mere existence of distinct hydrogen lines points to a pressure of less than half an atmosphere, as was shown by Evershed,[8] and the lines 4111, 4097, 3912 of chromium,[9] 3421, 3183 of barium,[10] and 4355, 4108, 3972 of calcium,[11] which are sharp and distinct in the solar spectrum, but which only lose their diffuseness in the laboratory under vacuum conditions, indicates pressures probably far lower than 0.1 atmospheres. The lines of doubly ionized nitrogen, which are seen as sharp clear absorption lines in the early B stars and the cooler O stars,[12] are also somewhat hazy under even the finest laboratory conditions,[13]

[7] St. John and Babcock, Ap. J., 60, 32, 1924.
[8] M. N. R. A. S., 82, 394, 1922.   [9] King, Ap. J., 41, 110, 1915.
[10] King, Ap. J., 48, 32, 1918.
[11] Saunders, quoted by Russell and Stewart, Ap. J., 59, 197, 1924.
[12] Payne, H. C. 256, 1924.
[13] A. Fowler, M. N. R. A. S., 80, 692, 1920.

EVIDENCE FROM LINE WIDTH 39

and probably arise in regions of very low pressure in the stellar atmosphere.

(c) *Widths of lines.* — The width of an absorption line, produced by "Rayleigh Scattering" close to resonance conditions, is given by Stewart [14] as

$$\Delta = (5.8 \times 10^{-13})\lambda\sqrt{N}$$

where Δ is the observed width of the line, λ the wave-length expressed in the same units, and N the number of molecules per square centimeter column in the line of sight.

It is unfortunate that the widths of Fraunhofer lines are hard to measure and difficult to interpret. Results obtained from objective prism spectra will probably differ from those derived with the aid of a slit spectrograph, and moreover, in estimating a line with wings it is hard to judge what should be regarded as the "true" line width. Russell and Stewart [15] estimate $\Delta/\lambda = 10^{-4}$ for the D lines in the solar spectrum. Then, on the assumption that the reversing layer has a thickness of a hundred kilometers, the partial pressure of neutral sodium in the reversing layer, as derived by Russell and Stewart from the formula just quoted, is of the order 10^{-9} atmospheres. At the solar temperature, 5600°, about 99 per cent of the sodium present is in the ionized condition,[16] and thus the total partial pressure of sodium atoms may be of the order 10^{-7} atmospheres. If it be assumed that sodium constitutes about 5 per cent of the total material present, the *total* pressure thus derived is of the order 10^{-6} atmospheres.

The D lines are of course the ultimate lines of neutral sodium. It will be shown [17] in Chapter IX that the partial electron pressure in the region from which ultimate lines originate is probably between 10^{-9} and 10^{-10} atmospheres at maximum. When 99 per cent of the atoms are ionized, the pressure rises by a factor of about 100, and the corresponding partial electron pressure becomes between 10^{-7} and 10^{-8} atmospheres. As the total pressure is probably, at the solar temperature, about twice the

[14] Stewart, Ap. J., in press.
[15] Ap. J., **59**, 197, 1924.
[16] Chapter VI, p. 99.
[17] Chapter IX, p. 137.

partial electron pressure, the total pressure should be nearer to 10^{-7} atmospheres.

The total pressure derived in Chapter IX is the pressure corresponding to the median frequency of the sodium atoms that send out light to the exterior — it may be regarded as the *average* pressure for the visible sodium. The total pressure derived from the line width, on the other hand, is the pressure at the *bottom* of the layer of visible sodium, and might therefore be expected slightly to exceed the average pressure for the visible sodium atoms. The difference encountered, 10^{-7} atmospheres for the average pressure, and 10^{-6} atmospheres for the total absorption pressure, is in the direction that would be anticipated, although it is larger than might have been expected. Neither value is, however, of very high accuracy, and probably the agreement can be regarded as quite satisfactory.

If the same formula be applied to the hydrogen lines, which may have a width [18] of the order of 5Å, high values for the partial pressure of hydrogen are obtained. The behavior of hydrogen in the spectra of the cooler stars,[19] and the abnormally high abundance [20] derived for it in Chapter XIII, suggest that here, again, a definite abnormality of the behavior of hydrogen is involved.

(d) *Flash Spectrum.* — It was pointed out by Russell and Stewart [21] that the density in the region that gives the flash spectrum must be exceedingly low. If this were not the case, the intensity of the scattered sunlight would be great enough, as compared to the flash itself, to register on the plate as continuous background in the time required to photograph the flash. The pressure thus estimated, from the minimum amount of material required to give scattered sunlight strong enough to be registered, is less than 2×10^{-5} atmospheres.

(e) *Radiative Equilibrium of the Outer Layers.* — At the edge of a star, where radiation pressure and gravitation no longer balance, and in consequence the existence of temperature and

---

[18] Shapley, H. B. 805, 1924.
[20] Chapter XIII, p. 188.
[19] Chapter V, p. 56.
[21] Ap. J., **59**, 197, 1924.

pressure gradients, such as we observe in the reversing layer, becomes possible, the equations given by Eddington [22] for the equilibrium of the interior no longer hold. The outer layers fall off more steeply than the equations predict, and in consequence it is not possible to use the equations in deriving values for the pressure or density corresponding to a layer near the boundary at a given temperature. It is certain, however, that the density deduced from the equations will be far too *high*, and so the predicted density at a given temperature may be used to indicate that the pressures at the boundary of a giant star are indeed very low.

The following table is adapted from the one given by Eddington for the relation between distance, r, from the center, density d, and temperature T, for a typical giant star of Class F7, effective temperature 6500°. The distance from the center is expressed in terms of the solar radius, the density in grams per cubic centimeter, and the temperature in absolute units. The last entry in the first column represents the total radius of the star.

r	d	T	r	d	T
0	0.1085	6,590,000°	4	0.0010	1,380,000
1	0.0678	5,640,000	5	0.00015	730,000
2	0.0215	3,840,000	6	0.0000093	290,000
3	0.0050	2,370,000	6.9	0.0000000	......

At a depth where the temperature is 290,000°, ten times the temperature in the reversing layer of any known star, the density given is about 10^{-5} grams per cubic centimeter. An atmosphere a hundred kilometers in thickness (the supposed approximate depth of the reversing layer) and of this density would contain only a hundred grams per square centimeter of surface. In order to bring the density into harmony with the densities derived for the reversing layer it is necessary to suppose that the value [23] of d falls to 0.4 per cent of its value at 290,000° as the

---

[22] Eddington, Zeit. f. Phys., 7, 371, 1921.
[23] Russell and Stewart (*loc. cit.*) show that there are about 0.4 grams of matter above the photosphere per square centimeter of surface.

temperature falls, from 290,000° to 29,000°, to 10 per cent of its value. The fall of density displayed in the table appears to be rapid enough to warrant this supposition; and in any case, as was pointed out earlier, the actual fall is probably greater than the formula predicts. The general theory of stellar equilibrium is, then, consistent with very low pressures in the reversing layer. More than this cannot be said, as the formulae are not directly applicable.

(f) *Observed Limit of the Balmer Series.* — The earlier members of the Balmer series of hydrogen are produced by the transfer of electrons from 2-quantum orbits to 3-quantum orbits (Hα), 4-quantum orbits (Hβ), and so forth. The later members of the series are associated with orbits of higher and higher quantum numbers. The major axis of the orbit varies as the square of the quantum number, and therefore a hydrogen atom which is producing, say, Hω, is effectively much larger than one which is giving rise to Hα. As was early suggested by Bohr,[24] the production of the higher members of the series must depend upon the possibility of existence of the corresponding outer orbits. As a preliminary assumption it appears probable that the existence of the larger orbits will depend on the proximity of neighboring atoms, and hence on the pressure.

The theoretical questions involved are very complex, and the present discussion is merely tentative. When the idea that the maximum number of lines that could be produced was a function of the pressure was first set forth, the available laboratory evidence appeared all to be in its favor. The maximum number of Balmer lines that had been produced in the vacuum tube was five, while it was well known that over twenty could be traced in absorption in some stellar atmospheres. Since that time, however, the work of R. W. Wood[25] has produced forty-seven lines of the Balmer absorption series of sodium in the laboratory at considerable pressures, and evidently the simple theory, relying on the mutual distances of the atoms to determine the

[24] Phil. Mag., **26**, 9, 1913.
[25] Phil. Mag., **37**, 456, 1919.

number of lines that can be produced, cannot be applied in this case. The matter has been discussed by Franck,[26] who points out that the outermost effective orbit in the sodium atom that gives the forty-seventh line must embrace large numbers of other atoms. He suggests that *collisions* are chiefly responsible for the production of the absorption lines.

Even though the simple theory is inapplicable to the laboratory conditions, it is not necessarily invalid in the stellar atmosphere, where conditions are far more simple, and where, in particular, the effects of collisions are negligible. There appears, moreover, to be a distinct observational correlation between the pressure and the number of observable hydrogen lines. The importance of the wave-length of the beginning of the continuous absorption, which lies just to the red of the last Balmer line observed, and extends toward the violet, was first indicated by Wright,[27] who recorded that the absorption head was farther to the red in α Lyrae than in α Cygni. This fact is obviously related to the difference in pressure in the atmospheres of the two stars, one of which is a normal A star, while the other is a supergiant. The observational and theoretical importance of the question has also been discussed by Saha,[28] and by Nicholson.[29]

The observational data in the hands of the writers just quoted were very meagre, and the present writer and Miss Howe [30] have recently attempted to obtain information on the number of observed Balmer lines in a large number of stars, and to examine the correlation with absolute magnitude. A distinct correlation is found between the number of lines observed and the reduced proper motion, which is chosen as the best available criterion of absolute magnitude for the numerous stars involved (Class A brighter than the fifth magnitude). It therefore appears that the pressure, and hence the proximity of the atoms, has some influence upon the possibility of the production of a

[26] Zeit. f. Phys., **1**, 1, 1920.
[27] Wright, Nature, **109**, 810, 1920.
[28] Saha, Nature, **114**, 155, 1924.
[29] Nicholson, M. N. R. A. S., **85**, 253, 1925.
[30] Unpublished.

line. The application of Bohr's original suggestion is hence of considerable interest, and the resulting pressures may profitably be compared with the pressures otherwise derived for the reversing layer.

The maximum number of lines seen, while quite consistent for plates made with the *same* dispersion, is somewhat increased when the dispersion is made much greater. The number of lines seen in the spectra of various stars with strong hydrogen lines, made with a dispersion of about 40 mm. between $H\beta$ and $H\epsilon$, varies between thirteen and twenty. The corresponding pressures, derived from Bohr's estimate that a pressure of about 0.02 mm. would be required for the production of thirty-three Balmer lines, and on the assumption that the pressure varies as the sixth power of the quantum number, lie between 10^{-3} and 10^{-4} atmospheres. These pressures are of course to be regarded as upper limits, for it is possible to miss several lines at the violet end of the series, and Wright, with larger dispersion, does indeed record twenty-four Balmer lines in α Cygni; on the other hand it is not likely that the estimated number will exceed the actual number of lines.

The pressures in the reversing layer, as derived from the observed limit of the Balmer series, are then of the same order as the pressures derived by the other methods outlined above. This is of especial interest because the method, if applicable, is a direct one, and gives results for individual stars, whereas all the other methods, excepting the one based on pressure shifts, are essentially indirect.

(g) *Ionization and Chemical Equilibrium.* — The evidence adduced by Russell and Stewart [31] has been greatly amplified by Fowler and Milne,[32] and by the data bearing on their theory which were subsequently published by the writer [33] and by Menzel.[34] It is not intended to present the evidence from ionization theory here in *support* of the low pressures inferred by the other

---

[31] Ap. J., **59**, 197, 1924.
[32] M. N. R. A. S., **83**, 403, 1923; **84**, 499, 1924.
[33] H. C. 256, 1924.
[34] H. C. 258, 1924.

SUMMARY OF DEDUCED PRESSURES

methods for the reversing layer. The pressure derived in the present chapter, and considered as independently established, will be used in Chapters VII ff. to derive a stellar temperature scale, for the reversing layer, from the line-intensity data presented.

SUMMARY

The following tabulation contains a synopsis of the reversing layer pressures derived by the methods that have been outlined.

Pressure Shifts	less than 10^{-1}
Line Sharpness	less than 10^{-1}
Line Width	10^{-6}
Flash Spectrum	2×10^{-5}
Radiative Equilibrium	order of 10^{-4}
Limit of Balmer Series	10^{-3} to 10^{-4} (upper limit)
(Ionization	10^{-4} to 10^{-9})

The extreme tenuity of the stellar atmosphere appears to be unquestionably established by the data set forth above, and a maximum effective pressure of 10^{-4} atmospheres may therefore be assumed in a discussion of the spectra of reversing layers.

CHAPTER IV

THE SOURCE AND COMPOSITION OF THE STELLAR SPECTRUM

THE spectrum of a laboratory source offers a somewhat inadequate comparison with the spectrum of a star. Matter can be studied terrestrially in small quantities only, and when a laboratory source is used in obtaining a spectrum, all the contributing material is collected into a very small region. With the stellar source it is quite otherwise. An enormous mass of matter, spread over a very large region, gives rise to the spectrum, and probably widely different physical conditions prevail at the origin of light of different wave-lengths. The present chapter contains a brief survey of the chief components which go to make up the stellar spectrum.

The spectrum of a star nearly always consists of a continuous background, in which the energy distribution corresponds more or less to that of a black body, and of absorption and emission lines and bands. The observed stellar spectrum is the integrated contribution from all parts of the disc, the unlined portion representing radiation that passes undisturbed from the photosphere through the reversing layer, and the light *within* any individual absorption line coming from the greatest depth in the reversing layer that can be penetrated by light of the corresponding frequency. This depth, which is a function of the monochromatic coefficient of absorption for the wave-length considered, is negligible when compared with the radius of the star.

DESCRIPTIVE DEFINITIONS

The solar atmosphere is probably qualitatively representative of all normal stellar atmospheres. It has been satisfactorily

DEFINITIONS

described by Russell and Stewart:[1] "At the top is a deep layer, the *chromosphere*, in which the gases are held up by radiation pressure, acting on individual atoms. The pressure and density in this layer increase slowly downwards (as gravity somewhat overbalances radiation pressure) and the pressure at its base may be of the order of 10^{-7} atmospheres, or 0.0001 mm. of mercury. Below this level, gravity is predominant in the equilibrium, and the pressure increases rapidly with depth — the temperature remaining nearly constant, and not far from 5000°, so long as the gases are transparent. This region is the *reversing layer*. When the pressure reaches 0.01 atmosphere, the general absorption by electron collisions begins to render the gas hazy. This opacity increases greatly with the pressure, and the reversing layer passes, by a fairly rapid transition, into the *photosphere*, which on the scale on which we have to study it resembles an opaque mass. As soon as the opacity becomes important the temperature rises in accordance with the theory of radiative equilibrium developed by Schwarzschild and Eddington. The observed effective photospheric temperature is a mean value for the layers from which radiation escapes to us."

The photosphere, as has been stated, is at an extremely small depth compared with the radius of the star. Taking the sun as an example, it is estimated by Russell and Stewart[2] that the reversing layer, which, with the chromosphere, is responsible for all the solar phenomena that can be spectroscopically studied, consists of about four tenths of a gram of matter per square centimeter of surface, and is only a few hundred kilometers in thickness. As this embraces only about 10^{-11} of the mass and 10^{-9} of the volume of the sun, it is clear that the features that can be studied spectroscopically are purely superficial, and that the larger aspects of stellar composition and constitution are left essentially untouched.

[1] Ap. J., **59**, 197, 1924.
[2] *Ibid.*

THE CONTINUOUS BACKGROUND

The continuous background of the spectrum represents the photosphere — the deepest layers from which we receive light. The energy that produces it is practically the total energy output of the star. While the actual distribution of energy in the spectrum probably conforms, in general, to that of a black body, the *observed* distribution naturally deviates considerably. But when corrections have been applied for atmospheric absorption, the resulting energy curves so far obtained do not appear to furnish certain evidence of serious deviation from blackness, although several investigators have suggested that their measures lead to this conclusion.[3-5]

If it is admitted that the energy distribution in the continuous background is sensibly black, the application of the Planck and Wien formulae furnishes methods of deriving the effective temperatures of stars from the energy distribution and the position of maximum intensity, respectively. The energy curve has therefore been extensively studied, both photographically and photometrically, and our present knowledge of stellar temperatures rests primarily upon work of this nature. The solar spectrum has been the subject of exhaustive photometric researches by Abbot[6] and Wilsing,[7] and the theory of the energy distribution, and its relation to the law of darkening, have been discussed by Lindblad,[8] and by Milne.[9] In a discussion of the solar energy curve, Milne[10] shows that the continuous spectrum can be regarded as that of a black body displaced to the violet, and that the displacement can be ascribed to the distortion of a normal black body curve by the presence of strong absorption.

H. H. Plaskett,[11] in applying the wedge method of spectro-

[3] H. H. Plaskett, Pub. Dom. Ap. Obs., 2, 258, 1923.
[4] C. G. Abbot, Ap. J., 60, 87, 1924. [5] Baillaud, C. R., 178, 1604, 1923.
[6] The Sun, 1911. [7] Potsdam Pub., 66, 1913.
[8] Lindblad, Upps. Univ. Arsskr., 1, 1920.
[9] Milne, Phil. Trans., 223A, 201, 1922.
[10] Milne, M. N. R. A. S., 81, 362 and 381, 1921.
[11] H. H. Plaskett, Pub. Dom. Ap. Obs., 2, 213 1923.

photometry to the same problem, took care to measure continuous background intensities in spectral regions free from absorption lines stronger than 0 per Angstrom, as measured on Rowland's scale of intensities. In this way he obtained a series of measures which should give a distribution sensibly free from distortion. His result for the solar temperature agrees more nearly with that derived from the solar constant than do the results of previous observers, and therefore the idea that the continuous background approximates to blackness is borne out by observations made with the proper precautions. R. H. Fowler [12] has remarked that "there is no longer any large discrepancy between the solar constant and the color temperatures, and one may hope that further more accurate work will leave them in full agreement."

The position of maximum intensity governs the *color* of the star, which is quite unrelated to the colors absorbed and radiated by the atoms in the reversing layer. In some of the Wolf-Rayet stars, apparently at very high temperatures and with atmospheres under special conditions of excitation, the continuous spectrum appears extremely faint, although there seems to be no reason for supposing that this is not merely an effect of contrast with the powerful emission "bands." The writer believes that long exposures would demonstrate the presence of continuous background for all such stars.[13] In the spectra of some gaseous nebulae, however, no continuous background has as yet been observed,[14] nor would any be expected, if our conception of the tenuity of these bodies is correct, unless they shine partly by pure reflection. (For example, the presence of some reflected starlight is inferred from the existence of a continuous background for the Orion nebula.) The transparency of gaseous nebulae to the light of stars indicates that their general opacity is extremely low, and it is this general opacity that is operative in producing the continuous background of a photosphere.

[12] Observatory, 47, 160, 1924.
[13] H. C. 263, 1924.
[14] Hubble, Mt. W. Contr. 241, 1922.

THE REVERSING LAYER

The reversing layer, comprising the layers above the photosphere, where the general opacity has greatly decreased and selective opacity begins to be appreciable, is responsible for the lines in the spectrum, which form the major part of the material of stellar spectroscopy. When the energy flowing out through the reversing layer in any specified wave-length is *less* than the energy in the neighboring continuous background, an absorption line is produced in the spectrum.

Roughly speaking, if an atom absorbs the whole of the light of any given frequency that reaches it from below, it will re-emit all the energy so absorbed, and will in general do so in a random direction.[15] The intensity of the absorption line so formed will then be about 50 per cent of the intensity in the neighboring continuous background. This argument is merely illustrative; it must suffice to point out that if pure selective absorption is operative the spectrum will be crossed by lines that are considerably less intense than the background. If, on the other hand, the energy leaving the atmosphere with any wave-length is *greater* than the energy in the neighboring continuous background, a bright line or "emission" line appears in the spectrum. Actually, of course, it is no more an emission line than is an ordinary Fraunhofer line, for the difference between stellar absorption and emission is merely a matter of contrast with the continuous background. Both kinds of line are "full of light."

ABSORPTION LINES

The absorption lines vary greatly among themselves and from star to star, both in intensity and in general appearance. The metallic lines, more particularly those of ionized atoms, are often extremely narrow and sharp — a feature difficult to reproduce in the laboratory, and referable to the very low pressures in the stellar atmosphere.[16] Other lines, such as those of

[15] Milne, M. N. R. A. S., 84, 354, 1924.   [16] Chapter III, p. 38.

the Balmer series of hydrogen, may be of considerable width, and spread out into wings that extend as much as thirty Angstrom units on each side of the center of the line. Many other lines are probably winged, but are not of sufficient strength for the feature to be seen. The form of the wings and the general shape of the line are of high significance, and should ultimately give much information bearing on the structure of the stellar atmosphere.

Although the absorption lines are commonly regarded as "dark," the foregoing section indicates that they should always have an appreciable intensity even at their centers. Measures of the central intensities of strong absorption lines have been published by various investigators, and the results are not all in agreement. Schwarzschild[17] gives from a single measurement of the H and K lines in the solar spectrum (center of disc) with the Hartmann microphotometer, wings ten Angstrom units in width on either side of the line center, and a weakening of the intensity of the light, from the continuous background to the center of the line, of about two and a half magnitudes. Bottlinger's curves[18] appear to lead to considerable intensities at the centers of the hydrogen lines in the A stars. Others have suggested that the central intensities are considerably lower. Abbot[19] quotes estimates ranging from one fifth to one tenth of the continuous background for solar lines, and H. H. Plaskett[20] states that the *faintest* stellar lines have about one tenth the intensity of the continuous background, as measured by his wedge method.[21]

Determinations of central intensity by means of precise photometry have been made by Kohlschütter[22] and by Shapley,[23] objective prism spectra being used in both cases. Kohlschütter gives the results of the analysis of the spectra of twenty-one

[17] Sitz. d. Pr. Ak. d. Wiss., 47, 1183, 1914.
[18] A. N., 195, 117, 1913.
[19] The Sun, 251, 1911.
[20] Pub. Dom. Ap. Obs., 1, 325, 1922.
[21] Pub. Dom. Ap. Obs., 2, 213, 1923.
[22] A. N., 220, 326, 1924. [23] H. B. 805, 1924.

stars of Classes A and F by means of the Hartmann microphotometer. The darkening from the continuous background to the
center of the line is tabulated in his paper for $H\delta$, $H\epsilon+H$, K, $H\zeta$,
and $H\eta$; it ranges for $H\delta$ from 1.14 magnitudes for α Lyrae to
0.42 magnitudes for α Cygni. The corresponding central intensities are 35 and 68 per cent of the intensity of the continuous background. The method used by Shapley employed a
special set of apertures to obtain a graded series of images for a
comparison of the central intensity with that of the adjacent
background. Although it is not certain that all the very complex photographic and photometric difficulties involved were
overcome by this method, its results are presumably entitled to
greater weight than any other determinations of central intensity hitherto made. The intensity in the hydrogen absorption
lines of Vega was ascertained to be about 25 per cent of that of
the background.

SATURATION OF ABSORPTION LINES

The discussion outlined above presupposes that the substance
producing the absorption line in the reversing layer is present in
quantities great enough to absorb all the light of the appropriate
wave-length, subsequently re-emitting it and giving rise to an
absorption line with considerable central intensity. If the atom
in question is present in quantities too small for complete absorption to take place, the central intensity of the line produced
will of course be higher still. Such atoms are designated " unsaturated." Saturation has been described by Russell [24] as follows: — " For the strong lines . . . the absorption in the reversing layer is so great that a large increase in the number of
absorbing atoms present alters the strength of the line very little. For the weak components . . . absorption under ordinary
conditions is incomplete, and the strengthening (in the spectra
of sunspots) is noteworthy " — an increase in the amount of
available material produces an increase in the strength of the

[24] Am. Ast. Soc. Rep., 190, 1923.

line. The strong components are saturated, the weak ones are not. It should be noted that here there is an excess of *atoms* for the radiation. "Saturation" is used in another sense when the word is applied to the conditions at the center of a star,[25] where there is an excess of *radiation* for the atoms present.

EMISSION LINES

The emission lines observed in stellar spectra differ more widely among themselves than do the absorption lines, and theory has so far been less successful in suggesting the physical conditions under which they may arise.[26] The appearance of the bright-line flash spectrum of the sun, from a region that gives no appreciable continuous spectrum, is of interest in comparing emission and absorption lines. It is fairly obvious that if the source of the flash spectrum had the photosphere behind it, the bright line would appear as absorption lines — which is indeed the case when the sun is ordinarily observed. Russell assigns both the Fraunhofer lines and part of the flash spectrum to the same region, namely the upper reversing layer. The high-level flash is, of course, assigned to the lower chromosphere. The difference between absorption and *narrow* emission is, as was pointed out in an earlier paragraph, purely a matter of contrast. There has, however, been no satisfactory explanation of how the phenomenon displayed by an ordinary emission line can be produced — an atom that re-emits in some wave-length more light than it receives in that wave-length. Some form of "fluorescent" emission would seem to be involved, and the question is evidently an important one for spectrum theory.

The chief types of emission are found in (*a*) the long period variables at maximum, (*b*) the emission B stars, (*c*) the O stars, including the Wolf-Rayet stars. All these stars are apparently very luminous.[27] Emission is also found in some late dwarfs — for example the H and K lines are reversed in the spectrum of

[25] Eddington, Ap. J., **48**, 205, 1918.
[26] M. C. Johnson, M. N. R. A. S., **85**, 56, 1924.
[27] *Ibid.*

61 Cygni,[28] and doubly reversed in the solar spectrum. Furthermore the spectra of gaseous nebulae are almost entirely composed of emission lines; and completely abnormal types of stars, with spectra partly or wholly composed of emission lines, might also be mentioned, notably the novae,[29] η Carinae,[30] Merrill's "iron star,"[31] Z Andromedae,[32] and [33] B. D.+11°4673. The conditions under which bright lines appear vary so widely that a single theory is manifestly inadequate to account for the phenomenon in every case.

[28] Adams and Joy, Pub. A. S. P., 36, 142, 1924.
[29] Chapter V, p. 64.
[30] H. A., 28, 175, 1901.
[31] Pub. A. S. P., 36, 225, 1924.
[32] Br. A. Rep., 1924.
[33] A. J. Cannon, H. B. 762, 1924.

CHAPTER V

ELEMENTS AND COMPOUNDS IN STELLAR ATMOSPHERES

THE identification of stellar lines and bands with those observed in the laboratory has furnished a rich source of data for astrophysics. About 25 per cent of the observed solar lines are assigned to elements in Rowland's Table of Solar Spectrum Wave-lengths. The majority of the solar lines which are still unidentified are faint. Notwithstanding practical difficulties of identification caused by blending, and the consequent uncertainty of wave-length, most of the observed lines, at least in the cooler stars, have been satisfactorily accounted for. There remain some important strong lines and bands of unknown origin, which have been usefully summarized by Baxandall.[1]

The present chapter contains a summary of the stellar occurrence and astrophysical behavior of the chief spectrum lines which are of known origin and series relations. A few other lines, such as those of $C++$, $N++$, and $O+$ are included, as their series relations will probably be forthcoming in the near future. The observed chemical elements are arranged in order of atomic number. At the conclusion of the chapter the elements which have not been detected in stellar spectra are enumerated. The series notation employed follows the system advocated by Russell and Saunders,[2] which appears to meet, more fully than any other, the practical needs of modern spectroscopy.

HYDROGEN (1)

Hydrogen is represented in stellar spectra by the Balmer series ($2P-mD$); the ultimate lines are those of the Lyman series ($1S-mP$) in the far ultra-violet, and cannot therefore be

[1] M. N. R. A. S., **83**, 166, 1923; *ibid.*, **84**, 568, 1924.
[2] Ap. J., **61**, 38, 1925.

traced in the stellar spectrum. The occurrence of the secondary spectrum of hydrogen, ascribed to the hydrogen molecule H_2, has been suspected,[3] but not definitely established. Only one of the lines has been recorded, and this should almost certainly be attributed [4] to $N++$. The familiar Balmer series appear as emission lines in the Wolf-Rayet stars, but normally they are absorption lines in all succeeding classes.

The intensity of the hydrogen lines is at a maximum [5] in the neighborhood of Class Ao. They vary greatly in width, however, within a given spectral class,[6] and it is difficult to find a method of photometry applicable to the comparison of lines of very different widths. The maximum of the Balmer lines has been placed by Menzel [7] at A3. The writer is inclined to believe that no significant maximum can in fact be derived for the Balmer lines; beyond A5, however, their intensity falls off rapidly.

It is peculiar to the Balmer series to appear in every class of the normal stellar sequence, and its lines at maximum exceed in strength the lines of every other element which appears in stellar spectra, excepting those of ionized calcium.

Although hydrogen is presumably unable to give rise to an "enhanced" spectrum, as the atom only possesses one extra-nuclear electron, the lines of the Balmer series share with those of neutral helium the peculiarity of behaving like the lines of an ionized atom.[8] They are weakened in dwarf M stars, and greatly strengthened in the cooler super-giants, such as α Orionis. The peculiarity of the astrophysical behavior of the hydrogen atom also appears in the impossibly high value that is assigned by ionization theory to the relative abundance of this element.[9] An explanation, in terms of metastability, has been suggested by Russell and Compton,[10] but although the hypothesis ap-

[3] Wright, Lick Pub., 13, 242, 1918.
[4] A. Fowler, M. N. R. A. S., 80, 692, 1920.
[5] H. A., 91, 7, 1918.
[6] Fairfield, H. C. 264, 1924.
[7] H. C. 258, 1924.
[8] Ibid.
[9] Payne, Proc. N. Ac. Sci., 11, 192, 1925; Chapter XIII, p. 188.
[10] Nature, 114, 86, 1924.

HYDROGEN

pears very satisfactory in the case of hydrogen, it is not applicable to the similar problem of helium. Russell [11] has remarked that " there seems to be a real tendency for lines, for which both the ionization and excitation potentials are large, to be much stronger than the elementary theory would indicate."

The hydrogen lines are often conspicuously winged. Measures of the width and intensity-distribution of the wings are discussed elsewhere.[12] Wings are probably not peculiar to the hydrogen lines, but the hydrogen wings can be studied because of their strength. The feature is also seen in helium, calcium and iron lines, and wings of greater or less strength are probably universal.

The width of the hydrogen lines in A stars has been correlated with absolute magnitude, and used for the estimation of luminosities.[13] It appears, however, that the line width may not furnish an accurate measure of absolute magnitude, although it serves to discriminate stars having the c-character from those of smaller luminosity.[14] The occurrence of wings seems, moreover, to be independent of line width and of absolute magnitude.[15] These questions are connected with the problem of classifying the A stars, and are discussed in a later chapter.[16]

The continuous spectrum of hydrogen, beyond the limit of the Balmer series, corresponding to the continuous radiation observed in the laboratory for sodium by Wood,[17] and for helium by Lyman,[18] was first noted in stellar spectra by Sir William Huggins.[19] The beginning of the band appears just to the red of the last Balmer line observed.[20] It appears, from work in progress at the Harvard Observatory,[21] that the limit is nearer to the violet, the higher the luminosity, and in a nebular spectrum quoted by Hubble,[22] it almost coincides with the theoretical limit of the series.

[11] Personal letter.
[13] Mt. W. Contr. 262, 1922.
[15] Lindblad, Ap. J., 59, 305, 1924.
[17] Ap. J., 29, 100, 1909.
[19] Atlas, p. 85, 1892.
[21] Chapter III, p. 43.

[12] Chapter IV, p. 51.
[14] Fairfield, H. C. 264, 1924.
[16] Chapter XII, p. 168.
[18] Ap. J., 60, 1, 1924.
[20] Wright, Nature, 109, 810, 1920.
[22] Pub. A. S. P., 32, 155, 1920.

The largest number of hydrogen lines recorded is thirty-five, measured by Mitchell [23] in the flash spectrum. Thirty-three were observed in emission by Evershed [24] in the solar chromosphere, and Deslandres [25] traced twenty-nine in the spectrum of a bright solar prominence. Twenty-seven Balmer lines have been observed by Curtiss [26] in the spectrum of ζ Tauri — the greatest number recorded for the spectrum of a star. The number of Balmer lines observed is related in Chapter III to the pressure in the reversing layer.

HELIUM (2)

Helium is represented in stellar spectra by the $1^2S - m^2P$, $1^2P - m^2S$, $1^2P - m^2D$, $1S - mP$, $1P - mD$, and possibly the $1P - mS$ series. Lines associated with these series appear almost simultaneously as we progress through the O star sequence, attain a maximum [27] at B3, and have disappeared [28] in normal A stars. The ultimate lines are the $oS - mP$ series,[29] in the far ultra-violet, and cannot be traced in the stars.

The helium lines vary much in width and definition and are often winged. Their intensity does not certainly appear to vary with absolute magnitude within a given spectral class, and they cannot therefore be used in the estimation of spectroscopic parallaxes.[30] The question of absolute magnitude effects cannot be usefully pursued in the absence of more reliable parallaxes, for the B stars, than are at present available.

Although the lines of helium do not appear in the normal A star, they are observed in the spectrum of the super-giant α Cygni, where the pressure is presumably exceedingly low. The $1^2P - m^2D$ lines also appear in the flash spectrum.[31]

[23] Mitchell, Ap. J., **38**, 431, 1913.
[24] Phil. Trans., **197A**, 381, 1901.
[25] C. R., **114**, 578, 1892.
[26] Pub. Obs. Mich., **3**, 256, 1923.
[27] Payne, H. C. 256, 1924; *ibid.*, 263, 1924.
[28] Henry Draper Catalogue; criterion of class.
[29] Lyman, Phys. Rev., **21**, 202, 1923.
[30] Payne, Nature, **113**, 783, 1924.
[31] Mitchell, Ap. J., **38**, 407, 1913.

IONIZED HELIUM

The lines of ionized helium appear only in the hottest stars, being peculiar to the O sequence. The $4F-mG$ lines (the "Pickering," or "ζ Puppis" series) are well marked in the hotter O stars, although all the lines usually available are probably blended.[32] The alternate Pickering lines are practically superposed on the Balmer lines, and the components were separated for several stars of Class O by H. H. Plaskett.[33] The "4686" series ($3D-mF$) appears in absorption in all the so-called "absorption O stars," and is even faintly seen in some B0 stars. The line at 4686 appears very readily as an emission line, and the wide bright "band" at this wave-length, which is a conspicuous feature of the Wolf-Rayet stars, of gaseous nebulae, and of certain stages of a nova, is also presumably due to ionized helium.

LITHIUM (3)

The element lithium is represented in the sunspot spectrum by the 1^2S-m^2P (ultimate) doublet at 6707, which is not, however, strong enough to be detected in stellar spectra. Russell[34] has called attention to the fact that this line is fainter, in the sun, than would be anticipated from the terrestrial abundance of the element. Compton[35] has suggested that the faintness may be ascribed to low atomic weight, and the consequent blurring of the line by a Doppler effect, owing to the high velocity of thermal agitation.

CARBON (6)

There is no evidence of the presence of neutral carbon in stellar atmospheres. The apparent absence of the element is partly due to the fact that the ultimate[36] line is at 2478, too far in the ultra-violet to be detected. The spectrum of neutral carbon is as yet unclassified, and other lines cannot, therefore, be sought for in the stellar spectrum. The temperature at which the ele-

[32] H. C. 263, 1924.
[34] Mt. W. Contr. 236, 1922.
[36] De Gramont, C. R., 171, 1106, 1920.
[33] Pub. Dom. Ap. Obs., 1, 335, 1922.
[35] Mt. W. Contr. p. 160, 236, 1922.

ment vaporizes is given by Kohn and Guckel [37] as 4000°, and by Violle [38] as 3800°. The heat of vaporization has been evaluated by de Forcrand.[39] At stellar temperatures, the carbon present is probably vaporized, but possibly it is largely in combination as cyanogen or as an oxide, since spectra associated with these compounds appear in low-temperature stars.

IONIZED CARBON

Ionized carbon [40] is represented in the stellar spectrum by the fundamental doublet $(2^2D - m^2F)$, at 4267, and by the principal doublet $(3^2S - m^2P)$ at 6580. The occurrence of these lines is of great interest. The line at 4267 is found in the O stars, [41] reaches a maximum at B3, and is last seen [42] at B9. It also occurs in the spectra of some gaseous nebulae.[43] In the stellar spectra in which it occurs, the line is sharp and clear, and, apart from appearing as an emission line in certain stars of Class O, it has no abnormal stellar behavior.

The principal doublet at 6580 has been said to occur in the Wolf-Rayet spectrum,[44-46] and to be much stronger than the fundamental doublet. The identification has been discussed by Wright,[47] and does not appear to be very probable. A knowledge of the behavior of the line at 6580 in the late O and early B stars is greatly to be desired.

DOUBLY IONIZED CARBON

Merton [48] has described a spectrum, produced under conditions of high excitation, which shows several correspondences with the emission bands of the Wolf-Rayet stars. His spectrum

[37] Naturwiss., 12, 139, 1924.   [38] Quoted by de Forcrand.
[39] C. R., 178, 1868, 1924.
[40] A. Fowler, Proc. Roy. Soc., 105A, 299, 1924.
[41] H. H. Plaskett, Pub. Dom. Ap. Obs., 1, 351, 1922.
[42] Payne, H. C. 256, 1924.   [43] Wright, Lick Pub., 13, 193, 1918.
[44] W. W. Campbell, Ast. and Ap., 13, 448, 1894.
[45] J. S. Plaskett, Pub. Dom. Ap. Obs., 2, 287, 1924.
[46] T. R. Merton, Proc. Roy. Soc., 91A, 498, 1915.
[47] Wright, Lick Pub., 13, 193, 1918.
[48] T. R. Merton, Proc. Roy. Soc., 91A, 498, 1915.

COMPOUNDS OF CARBON 61

contains the fundamental and principal doublets of C+, as well as a number of other lines, which have not as yet been assigned to series. Some of these lines are probably to be referred [49] to the atom of C++, and the writer [50] considers it unnecessary to assume the occurrence of a higher degree of excitation for the Wolf-Rayet spectrum. Some of the lines which are bright in the spectra of emission-line stars have been attributed to C+++ on astrophysical grounds,[51] and also from a discussion of frequency differences.[52] The four strongest groups in Merton's spectrum, however, consist of triplets, and this points more probably to C++, as does also the ionization potential deduced astrophysically [53] from the behavior of the only group accessible in ordinary stellar spectra. When the doublets due to C+, and the triplets already mentioned, are accounted for in Merton's spectrum, there remain only two lines at 5696 and 5592. A line [54] with the latter wave-length is attributed by Fowler and Brooksbank [55] to O++. The evidence for stellar C+++ appears, therefore, to be inconclusive.

COMPOUNDS OF CARBON

Cyanogen. The bands headed at 3885, 4215, have been attributed [56] to the CN or the C_2N_2 radical, or to the molecule of nitrogen. The assignment to a particular atom is essentially a question for the terrestrial physicist, and to discuss it here would be out of place.[57] The bands are universally known as the "cyanogen bands," and this designation will therefore be adopted.

The 3885 and 4215 bands are conspicuous in G and K stars of low density,[58] and furnish a valuable method for the measure-

[49] A. Fowler, Proc. Roy. Soc., **105A**, 299, 1924.   [50] H. C. 263, 1924.
[51] R. H. Fowler and Milne, M. N. R. A. S., **84**, 502, 1924.
[52] D. R. Hartree, Proc. Camb. Phil. Soc., **22**, 409, 1924.
[53] R. H. Fowler and Milne, M. N. R. A. S., **84**, 502, 1924.
[54] M. N. R. A. S., **77**, 511, 1917.   [55] Wright, Lick Pub., **13**, 193, 1918.
[56] Mulliken, Nature, **114**, 858, 1924; Birge, Phys. Rev. **23**, 294, 1924; Freundlich and Hocheim, Zeit. f. Phys., **26**, 102, 1924.
[57] Kayser, Handbuch der Spektroskopie, Vol. VII, 132, 1924.
[58] Evershed, Kod. Bul. 36, 1913.

62 ELEMENTS IN STELLAR ATMOSPHERES

ment of absolute magnitude — a method which has been used both at Mount Wilson [59] and at Harvard.[60] The band at 3885 is largely responsible for cutting off the ultra-violet light of the cooler stars.[61]

Cyanogen absorption has been reported as early [62] as A0, and according to Lindblad [63] it reaches a maximum at K2. The cyanogen bands reach great intensity in the N stars, and are indeed the most conspicuous feature of these spectra. Shane [64] places the maximum in Class N, and is doubtless correct in so doing. The maximum given by Lindblad refers to the series G K M, and the N stars are notoriously not members of that sequence.[65] Cyanogen is also a typical constituent of the comet-head spectrum.[66–68]

Carbon Oxides. The band spectrum attributed to the CO molecule,[69, 70] is a strong feature of the spectra of N and R stars.[71, 72] It is also the chief component of the spectrum of the comet tail, which has been reproduced in the laboratory, at very low pressures, by A. Fowler.[73]

Swan Spectrum. The bands of the Swan spectrum are clearly to be assigned to some compound of carbon,[74, 75] but the source is not as yet certainly established. They are characteristic of the comet head.[76] Another band, presumably to be associated with the Swan spectrum, was identified in the heads of nine

[59] Lindblad, Mt. W. Contr. 228, 1922.
[60] Shapley and Lindblad, H. C. 228, 1921.
[61] Lindblad, Mt. W. Contr. 228, 1922.   [62] Shapley, H. B. 805, 1924.
[63] Mt. W. Contr. 228, 1922.   [64] L. O. B. 329, 1919.
[65] Rufus, Pub. Obs. Mich., **3**, 257, 1923.
[66] A. Fowler, M. N. R. A. S., **70**, 176, 1909.
[67] Evershed, M. N. R. A. S., **68**, 16, 1907.
[68] Pluvinel and Baldet, Ap. J., **34**, 89, 1907.
[69] Merton and Johnson, Proc. Roy. Soc., 103A, 383, 1923.
[70] A. Fowler, M. N. R. A. S., **70**, 176, 1909.
[71] Hale, Ellerman, and Parkhurst, Yerkes Pub., **2**, 253, 1903.
[72] Shane, L. O. B. 329, 1919.
[73] M. N. R. A. S., **70**, 176 and 484, 1909.
[74] Strutt, Proc. Phys. Soc. Lond., **23**, 147, 1911.
[75] Stead, Phil. Mag., **22**, 727, 1911.
[76] A. Fowler, M. N. R. A. S., **70**, 176 and 484, 1909.

NITROGEN 63

comets by Baldet.[77] The ordinary Swan bands are also identified in the comet tail.[78]

Hydrocarbon. The identity of the " G " band with the 4314 hydrocarbon group was pointed out by Newall, Baxandall and Butler.[79] The strength of the band is increased, in the stellar spectrum, by the superposition of the $1^3P - m^3P'$ lines of calcium, and by the $1^5F - 1^5D$ lines of titanium, as well as other metallic lines, but the presence of the hydrocarbon band is certain, and is of the highest interest. The " G " band is first seen in some spectra [80] of Class A, and it attains a maximum at G or K.

The number of carbon compounds which occur lends plausibility to the suggestion that much of the stellar carbon is in combination at temperatures below 5000°.

NITROGEN (7)

The spectrum of neutral nitrogen has not as yet been satisfactorily analyzed into series.[81] It is quite possible that the first ionization that takes place is the ionization of the molecule,[82] which is accompanied by the production of the well known band spectrum. This spectrum has not been observed in the stars; presumably it would appear at lower temperatures than those involved in the coolest spectral classes. It is, however, stated to be a conspicuous feature of the spectrum of the aurora,[83, 84] and it is found in the spectrum of the comet head.[85] These occurrences seem to point to very low temperature and pressure at the source. It is possible that much of the nitrogen present in cooler stars is in combination with carbon.[86]

The green Aurora line was thought by Vegard [87] to coincide

[77] C. R., 177, 1205, 1923.
[78] Pluvinel and Baldet, Ap. J., 34, 89, 1911.
[79] M. N. R. A. S., 76, 640, 1916.   [80] Chapter XIV, p. 167.
[81] Fowler, Report on Series in Line Spectra, 164, 1922.
[82] Smyth, Proc. Roy. Soc., 103A, 121, 1923.
[83] Vegard, Videns. Skr., 1, nos. 8, 9, 10, 1923, where previous work is summarized.
[84] Vegard, Proc. Amst. Ac., 27, 1 and 2, 1924.
[85] A. Fowler, M. N. R. A. S., 70, 484, 1909.
[86] See p. 61.   [87] Vegard, Proc. Amst. Ac., 27, 1 and 2, 1924.

with a line emitted in the laboratory by solid nitrogen. The conclusion was questioned by McLennan and Shrum,[88] who failed to produce the line under similar conditions, and subsequently found a line, of the same wave-length as the aurora line, in the spectrum of a mixture of oxygen and helium.[89] Various previous attempts to identify the aurora line with a line produced in the laboratory had failed conspicuously.[90]

IONIZED NITROGEN

The spectrum of ionized nitrogen has recently been analyzed by A. Fowler.[91] The line which is most conspicuous in stellar spectra is the one at 3995 (P−S), which appears[92] at B0 or earlier, reaches maximum at B5, and is last seen at A0. Many of the fainter lines[93] are not observed.

DOUBLY IONIZED NITROGEN

The lines of doubly ionized nitrogen were singled out by Lockyer[94] as showing "abnormal behavior" — they do not appear in the same classes as the N+ line. The early work on the subject is discussed by Baxandall.[95] The most conspicuous lines are those at 4097, 4103, and they attain great intensity in the O stars;[96] they are, for example, very conspicuous in 29 Canis Majoris. H. H. Plaskett[97] places the maximum of the N++ lines in the Victoria class O7. They are last seen in some B0 stars.

The occurrence and behavior of the N+ and more especially the N++ lines in the Nova spectrum has been the subject of numerous investigations.[98–103]

[88] Proc. Roy. Soc., 106A, 138, 1924.
[89] Nature, 115, 382, 1925.
[90] Lord Rayleigh, Proc. Roy. Soc., 100A, 367, 1921; ibid. 101A, 312, 1922.
[91] A. Fowler, Proc. Roy. Soc., 107A, 31, 1925.
[92] Payne, H. C. 256, 1924.
[93] Ruark, Mohler, Foote, and Chenault, Bur. Stan. Sci. Pap. 480, 1924.
[94] Proc. Roy. Soc., 82A, 532, 1909.
[95] Pub. Solar Phys. Comm., 1910.
[96] Payne, H. C. 256, 1924.
[97] H. H. Plaskett, Pub. Dom. Ap. Obs., 1, 356, 1922.
[98] Paddock, Pub. A. S. P., 31, 54, 1919.
[99] Plaskett, J. R. A. S. Can., 12, 350, 1918.
[100] Baxandall, Pub. A. S. P., 31, 297, 1919.
[101] Wright, M. N. R. A. S., 81, 181, 1920.
[102] Ibid. Pub. A. S. P., 32, 276, 1920.
[103] Stratton, M. N. R. A. S., 79, 366, 1919.

Oxygen (8)

The ultimate lines of neutral oxygen occur[104] at a wavelength of about 1300, and accordingly cannot be observed in the spectra of stars. It was long supposed that neutral oxygen was entirely absent, but the $1^5S - m^5P$ triplet at 7700 is observed in the solar spectrum,[105] is strengthened in sunspots, and is strong in the high level chromosphere.[106] The ionization and excitation potentials corresponding to the production of these lines are of the same order as those for the Balmer series of hydrogen, and the astrophysical behavior of the triplet should therefore be similar to that of the hydrogen lines, with a maximum at or near Ao. Special work in the red is, however, required to trace the behavior of the series. The second member, the triplet at 3947, is not certainly present in the solar spectrum, and is not recorded for any star of Class A. In the laboratory, the second triplet is about as powerful as the first,[107] and its apparent weakness at the theoretical maximum is difficult to explain.

Ionized Oxygen

The spectrum of ionized oxygen should consist of pairs, and numerous lines have been tabulated as belonging to this atom.[108] The lines are found in B stars, as seems first to have been noticed by Lunt.[109] According to the present writer,[110] they are first seen at B0, although H. H. Plaskett, working with slit spectra, records[111] some O+ lines in Class O. The maximum of the O+ lines falls between B1 and B2, and their disappearance is mentioned[112] as a criterion of Class B3.

The lines at 4069, 4072, 4076, appear to form a triplet, but are more probably two pairs with two of the lines coalesced. Some stronger lines (page 207) persist in Class B5.

---

[104] Hopfield, Ap. J., **59**, 114, 1924.
[105] Runge and Paschen, Wied. An., **61**, 641, 1897.
[106] Curtis and Burns, unpub.   [107] A. Fowler, Report, 167, 1922.
[108] *Ibid.*   [109] An. Cape Obs., **10**, 5B, 1906.
[110] Payne, H. C., 256, 1924.   [111] Pub. Dom. Ap. Obs., **1**, 325, 1922.
[112] Henry Draper Catalogue, H. A., **91**, 1918.

DOUBLY IONIZED OXYGEN

The spectrum of O++ has been tabulated by A. Fowler and Brooksbank,[113] but not analyzed into series. The lines of this atom are certainly present [114, 115] in the stars of Class O. The astrophysical behavior of the lines of doubly ionized oxygen has led to the estimation of an ionization potential [116, 117] of 45 volts for the corresponding atom.

COMPOUNDS OF OXYGEN

Oxides. — Numerous oxides, such as carbon monoxide CO, titanium oxide TiO_2, zirconium oxide ZrO_2, and water H_2O, are present in the cooler stars. The metallic oxides are discussed under the corresponding metallic element. The occurrence of steam in the spectrum of the sunspot was announced by Cortie,[118] who supported his argument, originally based upon the widening, over sunspots, of telluric water vapor bands, by the observation that the presence of water vapor is essential, in the laboratory, to the production of the spectrum of magnesium hydride, which also occurs in the sunspot spectrum.[119] It is possible that the formation of oxides may account for the weakness of the spectrum of neutral oxygen in the cooler stars, but this explanation can hardly account for the absence of the second member of the $1^5S - m^5P$ series from the spectra of the A stars, where the lines should have their maximum intensity.

Ozone. — The ozone bands which appear in solar and stellar spectra have been shown by Fowler and Strutt [120] to be of telluric origin. The maximum thermal formation of ozone occurs [121] at 10^{-7} atmospheres and 3500°, and thus its presence in giant K and M stars might possibly be anticipated.

[113] M. N. R. A. S., **77,** 511, 1917.
[114] Wright, Lick Pub., **13,** 193, 1918.
[115] H. H. Plaskett, Pub. Dom. Ap. Obs., **1,** 325, 1922.
[116] Payne, H. C. 256, 1924; Proc. N. Ac. Sci., **10,** 322, 1924.
[117] R. H. Fowler and Milne, M. N. R. A. S., **84,** 499, 1924.
[118] Cortie, Ap. J., **28,** 379, 1908.
[119] A. Fowler, M. N. R. A. S., **67,** 530, 1907.
[120] Proc. Roy. Soc., **93A,** 577, 1917.
[121] Riesenfeld and Beja, Medd. Vetens. Nobelinst., **6,** 8, 1923.

MAGNESIUM

SODIUM (11)

The ultimate lines ($1^2S - m^2P$) of the neutral atom of sodium are the D lines, which lie at 5889, 5895. These are the only sodium lines which are certainly identified in stellar spectra.[122] They are first seen in the later B classes, and appear to be strengthened in cool stars, in accordance with theory.

Stationary sodium lines are observed [123] in β Scorpii, δ Orionis, and other Class B stars.[124]

The D lines are said to show an absolute magnitude effect, being strengthened in giant stars.[125]

No lines of ionized sodium are found in stellar spectra, presumably because they all lie in the far ultra-violet.

MAGNESIUM (12)

The neutral atom of magnesium is represented in the solar spectrum by the $1P - mD$, the $1^3P - m^3D$, and the $1^3P - m^3S$ series, and the first triplet of the latter series constitutes the conspicuous "b" group in the green. The "b" group and the second member of the $1^3P - m^3D$ series, the triplet near 3800, are first seen [126] at A0, have a maximum near K2 or K5, and are still strong in the coolest stars examined. The $1S - m^3P$ series, represented in the solar spectrum by a line at 4571, are faint ultimate lines; the strong ultimate lines [127] are the $1S - mP$ lines beginning at 2852, and are therefore outside the range of observed solar and stellar spectra.

IONIZED MAGNESIUM

The ionized magnesium atom gives rise to the important combination doublet at 4481 ($2^2D - m^2F$). These lines appear in the O sequence,[128] reach maximum [129] at A2 (not at A0, as stated by several investigators), and are lost in the increasing strength of

---

[122] Menzel, H. C. 258, 1924.   [123] Heger, L. O. B. 326, 1918.
[124] Heger, L. O. B. 337, 1922.   [125] Luyten, Pub. A. S. P., **35**, 175, 1923.
[126] Menzel, H. C. 258, 1924.   [127] de Gramont, C. R., **171**, 1106, 1920.
[128] H. H. Plaskett, Pub. Som. Ap. Obs., **1**, 325, 1922.
[129] Menzel, H. C. 258, 1924.

the iron line at the same wave-length, at about F2. The doublet varies with absolute magnitude, and may be found to furnish a useful criterion of that quantity. It has been used by H. H. Plaskett [130] in the estimation of the temperatures of some of the stars of Class O.

COMPOUNDS OF MAGNESIUM

Magnesium hydride. — The compound magnesium hydride, MgH_2, which has been studied in the laboratory by Brooks [131] and Fowler,[132] was detected by the latter in the sunspot spectrum.[133] It is perhaps significant that the only other hydride reported in celestial spectra is that of calcium, the next heavier alkaline earth after magnesium.

ALUMINUM (13)

Neutral aluminum is represented in the solar spectrum by the $1^2P - m^2S$ lines, the series that constitutes the ultimate lines in the third column elements of the periodic table.[134] The two conspicuous lines of the series in aluminum are those at 3944, 3957, and they are strengthened in cool stars,[135] in accordance with theory. They are especially mentioned as being strong in the spectrum [136] of 61 Cygni, as might be expected for a dwarf star. The $1^2P - m^2D$ series is also traced in the solar spectrum, but is too far in the ultra-violet to be studied effectively in the stars.

The series lines [137] of Al+ and Al++, although they might be expected in the B stars, apparently have not yet been traced in stellar spectra.

SILICON (14)

Four stages of the silicon atom are observed in stellar spectra. The line at 3905 is found in the coolest stars, has an observed

---

[130] H. H. Plaskett, Pub. Dom. Ap. Obs., **1**, 325, 1922.
[131] Proc. Roy. Soc., **80A**, 218, 1907.
[132] Phil. Trans., **209A**, 447, 1909.   [133] M. N. R. A. S., **67**, 530, 1908.
[134] R. H. Fowler and Milne, M. N. R. A. S., **83**, 403, 1923.
[135] Menzel, H. C. 258, 1924; p. 122.
[136] Adams and Joy, Pub. A. S. P., **36**, 142, 1924.
[137] Paschen, An. d. Phys., **71**, 151, 1923.

SILICON

maximum [138] at G5, and disappears at about F0. This line is regarded by A. Fowler [139] as the ultimate line of the neutral atom, and on this basis an ionization potential of 10.6 volts was assigned to silicon. In view of the fact that the line appears to have a maximum within the stellar sequence, and is of temperature class II, according to King,[140] while the true ultimate line [141] of silicon is at 2881, it seems possible that 3905 is actually a subordinate line.

IONIZED SILICON

Ionized silicon is represented by the lines 4128, 4131, which appear at B0, attain maximum [142] at A0, and disappear at F0. These lines are of especial interest, as they form the characteristic feature of the "silicon stars" which occur in the early A classes. The silicon stars are specially discussed [143] in Chapter XII.

DOUBLY IONIZED SILICON

The lines associated with the atom Si++ which appear in stellar spectra are the three at the wave-lengths 4552, 4568, 4574. These lines are first seen at B0, have a maximum [144] between B1 and B2, and disappear at B3. Fowler [145] regards these lines as constituting a principal triplet; it might be expected, however, that principal lines would show a more persistent maximum.

TRIPLY IONIZED SILICON

The atom of silicon which has lost three electrons is the most highly ionized atom of which we have certain evidence in stellar spectra. The lines at 4089, 4096, and 4116 are strong [146] among the cooler O stars, and are last seen at Class B0. The hotter O stars, such as H.D. 165052, do not display the lines of Si+++, and probably the intensity of the lines has fallen, owing to the tem-

[138] Payne, H. C. 252, 1924.
[139] Bakerian Lecture, 1924.
[140] King, Pub. A. S. P., 33, 106, 1921.
[141] de Gramont, C. R., 171, 1106, 1920.
[142] Payne, H. C. 252, 1924.
[143] P. 169.
[144] Payne, H. C. 252, 1924.
[145] Bakerian Lecture, 1924.
[146] Payne, H. C. 263, 1924.

70   ELEMENTS IN STELLAR ATMOSPHERES

perature, which is above that required for the maximum of these lines.

SULPHUR (16)

The spectrum of neutral sulphur which has hitherto been analyzed is chiefly in the far ultra-violet,[147] and is therefore not traceable in the sun or stars.

Two sets of sulphur lines, differing in astrophysical behavior, were noted by Lockyer [148] at 4163, 4174, 4815, and at 4253, 4285, 4295. These lines have been attributed by the writer,[149] and by Fowler and Milne,[150] to S+ and S++ respectively. The S+ lines appear to be in pairs, and the S++ lines suggest a triplet, although one of the three lines is extremely faint in stellar spectra, and it would be expected that the once and twice ionized spectra of sulphur would display even and odd multiplicities respectively. The two series have maxima at B8 and at B1, but the stellar intensities of the lines are small. An amplification of our knowledge of stellar sulphur is greatly to be desired.

POTASSIUM (19)

The ultimate lines [151] of potassium (1^2S-m^2P) are at 7664 and 7699, and have been traced in the solar spectrum, although they are very faint. They appear to be absent from the flash spectrum.[152] Russell [153] expresses the opinion that they persist, with rising temperature, as far as F8 in the stellar sequence.

CALCIUM (20)

The element calcium is extensively represented in stellar spectra. The ultimate line of the neutral atom is at 4227 ($1S-mP$) and appears at A0. The line increases in strength in all cooler stars, in accordance with theory, and has a distinct variation with absolute magnitude. The $^3P-^3P'$, $^3P-^3D$ and $^3D-^3F$ mul-

[147] Hopfield, Nature, 112, 437, 1923.
[148] Proc. Roy. Soc., 80A, 50, 1907.
[149] H. C. 256, 1924.
[150] M. N. R. A. S., 84, 499, 1924.
[151] Fowler, Report on Series in Line Spectra, 1922.
[152] Curtis and Burns, unpub.   [153] Personal letter.

IONIZED CALCIUM 71

tiplets [154] are satisfactorily identified in the solar spectrum and can be traced with certainty in the spectra of stars cooler than Fo. The 1P−mS, 1P−mD, 1D−mF, and 1D−mP lines appear to be present with the appropriate intensities in the sun, but are too faint to be seen with small dispersion. Thus all the classified lines of calcium which are strong in laboratory spectra have been traced in the spectra of the sun and stars.

IONIZED CALCIUM

The H and K lines of ionized calcium are seen throughout the stellar sequence, and reach a maximum within the K type,[155] where their intensity is greater than that attained by any other line in any class. They vary with absolute magnitude.[156] In the sun the lines are doubly reversed, and they are probably singly reversed [157] in 61 Cygni.

Stationary calcium lines have long been known to occur in the spectra of certain spectroscopic binaries, having first been noticed by Hartmann [158] for δ Orionis. Various "calcium cloud" hypotheses have been advanced to account for the phenomenon. It appears, from several considerations, notably the apparent small oscillation of the calcium lines with the same period as the star, that there is some physical connection between the two. Lee [159] discussed the idea that the system of 9 Camelopardalis was surrounded by a cloud of calcium vapor, which, as he showed, could be made to account for the behavior of the lines of ionized calcium. The same idea was discussed by J. S. Plaskett, who suggested that we might "assume that the absorbing material is near to or envelopes the stars, which is probable from its wide distribution, and in this form it combines the two original hypotheses of interstellar and surrounding clouds." [160] The D

[154] Russell and Saunders, Ap. J., 61, 38, 1925.
[155] Menzel, H. C. 258, 1924.
[156] Ibid.
[157] Adams and Joy, Pub. A. S. P., 36, 142, 1924.
[158] Ap. J., 19, 268, 1904.
[159] Lee, Ap. J., 37, 1, 1913.
[160] Pub. Dom. Ap. Obs., 2, 16, 1924.

lines of sodium [161] and possibly the hydrogen lines [162] have been added to the list of stationary lines, and Plaskett [163] has suggested that the ultimate lines of the ionized atoms of strontium and barium should also show the effect, which has not yet, however, been observed.

SCANDIUM (21)

The element scandium [164] is represented in the solar spectrum by faint lines corresponding to the multiplets $1^2D - 5^2D'$, $1^2D - 4^2D'$, $1^2D - 2F'$. The multiplet $1^4F - 1^4F'$ may possibly be present, but the lines are very weak. The element is not recorded in the spectra of stars; most of the lines are unsuitably placed in the green.

IONIZED SCANDIUM

Six multiplets of ionized scandium, out of the eight tabulated by Meggers, Kiess, and Walters [165] appear in the solar spectrum, and all the corresponding lines have been traced in Rowland's tables. The intensity of two of the lines is great enough for their behavior to be traced through the stellar sequence, and they are greatly enhanced in the spectra of the c-stars. The ultimate lines are near 3600, but in the solar spectrum they are less powerful than the lines near 3500.

Table XI on page 73 contains, in successive columns, the series relations, the wave-length as determined in the laboratory, the intensity, the temperature class, and the attribution, solar intensity, and wave-length given by Rowland, for the six multiplets which lie within the observed range of the solar spectrum. Ultimate lines are designated by an asterisk.

TITANIUM (22)

The spectrum of titanium is so rich in lines, and is so largely represented in stellar spectra, that a tabulation would occupy

---

[161] Heger, L. O. B. 326, 1918; *ibid.* 337, 1922.
[162] Rufus, J. R. A. S. Can., **14**, 139, 1920.
[163] J. S. Plaskett, Pub. Dom. Ap. Obs., **2**, 344, 1924.
[164] Russell, Mt. W. Contr., in press.
[165] J. Op. Soc. Am., **9**, 355, 1924.

SCANDIUM

TABLE XI

Series	Wave-Length	Int.	Cl.	Attribution	Int.	Wave-Length
*$^3D_3-^3F_4$	3613.84	60	II	−, Sc	4	3613.947
$^3D_3-^3F_3$	3645.31	30	III	Sc?, −	3	3645.475
*$^3D_2-^3F_3$	3630.76	50	II		4	3630.876
$^3D_3-^3F_2$	3666.54	3	III		1	3666.676
$^3D_2-^3F_2$	3651.81	25	III	−, Sc	4	3651.940
*$^3D_1-^3F_2$	3642.79	40	II	Sc	2	3642.912
$^3D_3-^3D'_3$	3572.53	50	II	−, Sc	6	3572.71
$^3D_2-^3D'_3$	3558.55	20	II	(Fe	8	3558.672)
$^3D_3-^3D'_2$	3590.48	20	II		2	3590.609
$^3D_2-^3D'_2$	3576.35	35	II	−, Sc?	3	3576.527
$^3D_1-^3D'_2$	3567.70	20	II		4	3567.835
$^3D_2-^3D'_1$	3589.64	20	II		5	3589.773
$^3D_1-^3D'_1$	3580.94	30	II		5	3581.067
$^3P_2-^3P'_2$	5657.89	25	V E	Y, −	2	5658.09
$^3P_2-^3P'_1$	5684.21	12	V E		1	5684.415
$^3P_1-^3P'_2$	5640.99	15	V E		2	5641.206
$^3P_1-^3P'_1$	5667.16	9	V E		0	5667.368
$^3P_0-^3P'_1$	5658.35	8	V E		0	5658.561
$^3P_1-^3P'_0$	5669.05	10	V E		1	5669.258
$^3F_4-^3F'_4$	4374.46	40	III E	Sc, Fe?	3	3374.628
$^3F_4-^3F'_3$	4420.66	2			00	4420.832
$^3F_3-^3F'_4$	4354.60	5	V E		1	4354.776
$^3F_3-^3F'_3$	4400.38	30	III E	Sc	3	4400.555
$^3F_3-^3F'_2$	4431.35	3	V E		0	4431.525
$^3F_2-^3F'_3$	4384.80	6	IV E		0	4384.986
$^3F_2-^3F'_2$	4415.55	20	III E		2	4415.722
$^3F_4-^3D'_3$	4314.09	60	III E	Sc	3	4314.248
$^3F_3-^3D'_3$	4294.77	8	IV E	Zr	2	4294.932
$^3F_3-^3D'_2$	4320.73	50	III E	Sc	3	4320.90
$^3F_2-^3D'_3$	4279.95	1	−		−	− − − −
$^3F_2-^3D'_2$	4305.70	10	IV E		2	4305.871
$^3F_2-^3D'_1$	4325.00	40	III E	Sc	4	4325.152

an undue amount of space. From an examination of Rowland's tables of the solar spectrum, it appears that the fainter components of the multiplets invariably accompany the stronger ones, thus making the identifications certain. Only the stronger components are, however, powerful enough to appear in stellar spectra, with the dispersions ordinarily used.

The following multiplets, as analyzed by Russell[166] and Kiess,[167] are definitely present: $1^3F-1^3F'$, $1^3F-2^3F'$, $1^3F-1^3G'$, $2^3F-6^3G'$, $1^3F-1^3D'$, $1^3F-2^3D'$, $1^3P-2^3P'$, $1^5F-2^5G'$, $1^5F-2^5F'$, $1^5F-2^5D'$, $1^5P-1^5P'$. Doubtfully present are: $1^5P-4^5D'$, $2^3F-5^3F'$.

The maximum of these lines is difficult to determine; they are not well placed for measurement, many of the most important are seriously blended, and all are rather faint, even at maximum. They are first seen[168] at Class A2, and their maximum appears to be[169] at K2 or K5.

The solar intensities of the lines of both neutral and ionized titanium fall off regularly with increasing excitation potential. The subject is discussed in Chapter VII, as part of the evidence for the validity of the Saha theory.[170]

Ionized Titanium

The lines of ionized titanium are about as strong in the solar spectrum as those of the neutral atom. Many of them appear, with the lines of the ionized iron atom, with abnormal strength in the spectra of the c-stars.[171] The following multiplets[172] are present in the solar spectrum: $1^4F-1^4G'$, $1^4F-1^4F'$, $1^4F-1^4P'$, $2^4F-1^4G'$, $2^4F-1^4F'$, $2^4F-1^4P'$, $1^2F-1^4G'$, $1^2F-1^2F'$, $1^2F-1^2D'$, $1^2D-1^2F'$, $1^2D-1^2D'$, $1^2G-1^2F'$, $1^2G-1^2G'$, 1^4P-1^2D, $1^4P-1^4D'$, $1^2P-1^2D'$, $1^2P-1^4D'$, $1^2H-1^2G'$, $1^2H-2^2G'$, $2^2G-2^2G'$, $2^2G-2^2F'$, $2^2G-2^2G'$. Doubtfully present are $1^2P-1^2F'$, $2^2D-1^2F'$, $2^2D-1^2D'$, $2^2F-2^2G'$. The lines which are especially enhanced in the c-stars are: $2^2G-2^2F'$, $1^4P-1^2D'$, $1^2D-1^2F'$, $1^2G-1^2D'$, $1^2G-1^2F'$, $1^2P-1^2D'$, $1^2H-1^2G'$, $2^2P-1^2S'$.

The lines of ionized titanium come to a maximum at about Class F5, but a significant maximum is difficult to determine,

[166] Mt. W. Contr., in press. [167] J. Op. Soc. Am., 8, 609, 1924.
[168] Payne, Proc. N. Ac. Sci., 11, 192, 1925.
[169] Menzel, H. C. 258, 1924. [170] Chapter VII, p. 113.
[171] Maury, H. A., 28, 79, 1900. [172] Russell, Mt. W. Contr., in press.

for the lines are extremely sensitive to absolute magnitude. Menzel,[173] using β Cassiopeiae (classed by him as F0) for his typical star, found a maximum development of lines in that star. The present writer,[174] using the wider selection of stars enumerated in the appendix, obtains G0 as the maximum for Ti+. A glance at the measures [175] will indicate that the position of the maximum is in any case very uncertain, as the intensity does not change smoothly in going from class to class.

COMPOUNDS OF TITANIUM

The absorption bands of titanium oxide, TiO_2, are the characteristic flutings [176, 177] of the stars of Class M, and the strength of these bands has been proposed [178] as a criterion of class for the stars in which they are found. It is perhaps noteworthy that titanium, zirconium, and carbon, the only elements which give oxides in stellar spectra (hydrogen excepted) belong to the fourth group of the periodic system.

VANADIUM (23)

The vanadium lines are best identified by intensity from Rowland's table. The following multiplets [179] are present in the solar spectrum: $1^6D - 2^6D'$, $1^6D - 2^6F'$, $1^6D - 1^6P'$, $1^4D - 3^4F'$, $1^4F - 1^4F'$, $1^4F - 1^4G'$, $1^4F - 1^4G'$. The $1^6D - 2^6D'$ multiplet is well seen in stellar spectra from F0 onwards, and increases in strength as cooler stars are approached.[180] Slipher [181] called attention to the strength in o Ceti of the vanadium group near 4400, presumably the two multiplets $1^6D - 1^6P'$, $1^6D - 2^6F'$, with excitation potential 0.28 volts.

[173] H. C. 258, 1924.
[174] Chapter IX, p. 123.
[175] *loc. cit.*
[176] A. Fowler, Proc. Roy. Soc., **73A**, 219, 1904; M. N. R. A. S., **69**, 508, 1909.
[177] Hale, Adams and Gale, Ap. J., **24**, 185, 1906; Hale and Adams, *ibid.*, **25**, 75, 1907.
[178] Rep. of Spectral Class. Comm., I. A. U., 1925.
[179] W. F. Meggers, J. Wash. Ac. Sci., **13**, 317, 1923; *ibid.*, **14**, 151, 1924.
[180] Menzel, H. C. 258, 1924.
[181] Ap. J., **25**, 235, 1907.

IONIZED VANADIUM

Three multiplets, all far in the ultra-violet, are tabulated for ionized vanadium by Meggers, Kiess, and Walters,[182] and two of them are within the range of Rowland's table. All the lines of these, the $^3F-^3F$ and $^5F-^5G$ multiplets, have been satisfactorily identified with solar lines. The strength of the ultimate lines of ionized vanadium, which occur in the multiplet last named, is a little greater, in the solar spectrum, than that of the strongest lines of the neutral atom, at 4379, which are also ultimate lines.

The following tabulation contains, in the same form as Table XI, the data respecting the two multiplets which are identified in the solar spectrum.

TABLE XII

Series	Wave-Length	Int.	Cl.	Attribution	Int.	Wave-Length
$^3F_4-^3F'_4$	3727.348	20	–		1	3727.488
$^3F_3-^3F'_4$	3760.230	5	–		1	3760.364
$^3F_4-^3F'_3$	3718.163	3	–		oN	3718.291
$^3F_3-^3F'_3$	3750.873	15	–		2	3751.015
$^3F_2-^3F'_3$	3778.359	3	–	(Fe	3	3778.463)
$^3F_3-^3F'_2$	3743.63	3	–	(Cr	1	3743.726)
$^3F_2-^3F'_2$	3770.976	10			2	3771.116
*$^5F_5-^5G'_6$	3093.10	40	III Er		2N	3093.229
$^5F_5-^5G'_5$	3121.144	20	IV E	V	4	3121.270
*$^5F_4-^5G'_5$	3102.301	40	III Er	V	3	3102.404
$^5F_5-^5G'_4$	3145.35	–	–		3	3145.484
$^5F_4-^5G'_4$	3126.221	25	IV E	V, Fe	5	3126.319
*$^5F_3-^5G'_4$	3110.710	30	III Er	Ti, V	5Nd?	3110.810
$^5F_4-^5G'_3$	3145.979	5	V E	Zr	1	3146.091
$^5F_3-^5G'_3$	3130.270	25	III E	V	3	3130.380
*$^5F_2-^5G'_3$	3118.382	30	III Er	V	3	3118.498
$^5F_3-^5G'_2$	3145.344	10	IV E		3	3145.484
$^5F_2-^5G'_2$	3133.336	20	III E	V	2	3133.449
*$^5F_1-^5G'_2$	3125.286	40	III Er		5	3125.399

[182] J. Op. Soc. Am., **9**, 355, 1924.

CHROMIUM (24)

The lines of chromium were classified by Catalan,[183] and those which occur in the sun are comprised in the following multiplets:[184] 1^7S-1^7P, 1^7S-1^5P, 1^7S-2^7P, 1^5S-1^7P, $1^5S-1^5P'$, $^5S-1^5P$, 1^5D-1^5P, $1^5D-1^5P'$, 1^5D-1^5P, 1^7P-2^7S, 1^7P-1^7D, $^7P-2^7D$.

The ultimate lines 1^7S-1^7P, at 4254, 4274, 4289 increase with advancing type.[185] The maximum for subordinate lines [186] is at M1.

IONIZED CHROMIUM

Of the six multiplets of ionized chromium tabulated by Meggers, Kiess, and Walters,[187] only two are within the measured range of the solar spectrum, but every line in these two multiplets accords satisfactorily in wave-length and intensity with a

TABLE XIII

Series	Wave-Length	Int.	Attribution	Int.	Wave-Length
$^4D_1-^4P_2$	3328.34	3		2	3328.487
$^4D_1-^4P_1$	3336.33	5	Cr	2	3336.477
$^4D_2-^4P_3$	3324.06	3		4N	3324.19
$^4D_2-^4P_2$	3339.80	10	Co, Cr	3	3339.932
$^4D_2-^4P_1$	3347.83	6	Cr	3	3347.970
$^4D_3-^4P_3$	3342.58	10	Cr	3	3342.717
$^4D_3-^4P_2$	3358.50	10	Ti, Cr	4	3358.649
$^4D_4-^4P_3$	3368.04	20	Cr, -	5d?	3368.193
$^4D_4-^4F_5$	3132.04	20	-, Cr	4	3132.169
$^4D_3-^4F_4$	3124.97	20	Cr	4	3125.109
$^4D_4-^4F_4$	3147.22	5	Cr	3	3147.350
$^4D_2-^4F_3$	3120.36	15	Cr, -	3	3120.481
$^4D_3-^4F_3$	3136.69	5	Cr, Co	3	3136.822
$^4D_4-^4F_3$	3159.10	1		0	3159.225
$^4D_1-^4F_2$	3118.65	10	Cr, -	2	3118.764
$^4D_2-^4F_2$	3128.68	5	Cr, -	2	3128.819
$^4D_3-^4F_2$	3145.07	2		2	3145.251

[183] Catalan, An. Soc. Espan. Fis. y Quim., 21, 84, 1923.
[184] Russell, Mt. W. Contr., in press. [185] Chapter VIII, p. 124.
[186] Menzel, H. C. 258, 1924. [187] J. Op. Soc. Am., 9, 355, 1924.

line in Rowland's table. The ultimate lines are in the neighborhood of 2800, and are therefore unattainable. The lines, and the solar intensities, are contained in the appended table.

MANGANESE (25)

The lines of manganese are conspicuous in stellar spectra, and all the classified lines [188] within the range of Rowland's table are found in the solar spectrum, namely the multiplets $1^6S - 1^6P$, $1^6S - 2^6P$, $1^6D - 2^6P$, $1^6D - 6D'$, $1^8P - 2^8D$, $1^8P - 1^8S$, $1^4D - 1^4P$, $1^6P - 3^6D$, $1^6D - 1^6F$, $1^4D - 1^4F'$. The ultimate lines $1^6S - 1^6P$ are at 4030, and constitute a conspicuous group in the solar spectrum. They are well seen in the cooler stars, and are progressively strengthened with advancing type.[189] They first appear at A0. The $1^6D - 6D'$ multiplet, at 4018, 4041, 4055, 4084, etc., has a maximum, according to Menzel,[190] at K5.

IONIZED MANGANESE

Meggers, Kiess, and Walters [191] give one multiplet of ionized manganese, and this is within the range of Rowland's table. The multiplet was previously picked out by Catalan [192] as being analogous to the arc multiplet $1^6D - 2^6P$. All the lines can be satisfactorily identified with lines in the solar spectrum, as in the following table.

TABLE XIV

Series	Wave-Length	Int.	Cl.	Attribution	Int.	Wave-Length
$^5P_3 - ^5D_4$	3441.999	9	V	Mn	6	3442.118
$^5P_3 - ^5D_3$	3474.050	7	V	Mn	2	3474.197
$^5P_2 - ^5D_3$	3460.332	8	V	Mn, -	4d?	3460.460
$^5P_3 - ^5D_2$	3496.815	4	V	Co, Mn	3	3496.952
$^5P_2 - ^5D_3$	3482.918	7	V	Mn, -	5d	3483.047
$^5P_1 - ^5D_2$	3474.13	6	V	Mn	2	3474.287
$^5P_2 - ^5D_2$	3497.540	6	V	Mn	3	3497.668
$^5P_2 - ^5D_1$	3488.618	8	V	Mn	4	3488.817
$^5P_1 - ^5D_0$	3495.810	8	V	Mn	2	3495.974

[188] Catalan, Phil. Trans., 223A, 127, 1922.
[189] Chapter VIII, 124.
[190] H. C. 238, 1924.
[191] J. Op. Soc. Am., 9, 355, 1924.
[192] Phil. Trans., 223A, 127, 1922.

Iron (26)

The extensive occurrence of the arc lines of iron in the stellar spectrum is well known. The following multiplets [193] have been traced in the solar spectrum, and the corresponding lines are also to be traced in the spectra of the cooler stars: $1^5D - 1^7D'$, $1^5D - 1^5D'$, $1^5F - 1^5D'$, $1^5F - 1^5F'$, $1^5F - 1^3F'$, $1^5F - 2^5D'$, $1^3F - 1^3F'$, $1^3F - 2^5D'$, $1^3F - 2^5F'$, $1^3F - 1^5G$, $1^3F - 1^3G'$, $1^3F - 2^3F'$, $1^5P - 3^5D$, $1^5D - m^5F$, $1^5F - m^5F$, $1^7F - m^7D$. The iron lines have, in general,[194] a maximum at K2, but the only ultimate lines which are well shown in stellar spectra, the $1^5D - 1^7F$ lines near 4480, increase with advancing type to the end of the sequence.[195]

The following lines are used as criteria of absolute magnitude by Harper and Young:[196] 4202, 4250, 4272 ($1^3F - 1^3G$), 4072 ($1^3F - 2^3F'$), 4482 ($1^5D - 1^7F$).

Ionized Iron

The lines of ionized iron are strong in F stars of high luminosity, and are especially conspicuous in the stars which have the c-character. Menzel [197] places the maximum at A7, and the writer [198] finds it at F5. The following multiplets, as classified by Russell,[199] occur in the solar spectrum: $1^4G - 1^4F'$, $1^4G - 1^4D'$, $2^4F - 1^4F'$, $2^4F - 1^4D'$, $2^4F - 1^4P'$, $2^4P - 1^4F'$, $2^4P - 1^4D'$, $2^4P - 1^4P'$, $1^4P - 1^4F'$, $1^4P - 1^4P'$, $1^4D - 1^4F'$, $1^4D - 1^4D'$, $1^4D - 1^4P'$, $1^4F - 1^4F'$, $1^4F - 1^4D'$, $1^4F - 1^4P'$, $1S - 1^4F'$, $1S - 1^4D'$, $1S - 1^4P'$, $1^4F - 1^4G$, $1^4F - 1^4D$. Doubtfully present are: $1^6D - 1^4F'$, $1^6D - 1^4P'$, $1^6D - 1^4D'$. The ionized iron lines are strengthened, as are other enhanced lines, over sunspots, and many of the fainter components of multiplets are observed only in the spot spectrum.

[193] Walters, J. Op. Soc. Am., **8**, 245, 1924.
[194] Chapter VIII, p. 125. [195] *Ibid.*
[196] Pub. Dom. Ap. Obs., **3**, 7, 1925.
[197] H. C. 258, 1924.
[198] Chapter VIII, p. 126.
[199] Mt. W. Contr., in press.

COBALT (27)

The series relations for the arc spectrum of cobalt [200] have been published by Walters. Cobalt lines are frequent in the solar spectrum, but as the strongest of them lie near 3500, they cannot be traced in the spectra of stars. The following multiplets are certainly identified in the spectrum of the sun: $^4F-^4D$, $^4F-^4D'$, $^4F-^4F''$, $^4F-^4G$, $^4F'-^4D$, $^4F'-^4D'$, $^4F'-^4F''$, $^4F'-^4G$, $^4P-^4D$, $^4P'-^4D$, and $^4P-^4D'$. The incompletely observed multiplet $^4P'-^4D'$ is apparently absent from the solar spectrum.

NICKEL (28)

The series relations for nickel are as yet unpublished. The lines appear in great numbers in the solar spectrum, but they are not strong enough to be conspicuous in the spectra of the stars. The line 5476 appears to have a maximum at G0, indicating either that it is an enhanced line of nickel, or that it is blended with the enhanced line of some other element. The lines 5081, 4714 are strengthened in low temperature stars, and are probably due to neutral nickel. From the solar behavior of the lines of this element,[201] the ionization potential seems to be of the same order as that for cobalt, probably about 8 volts.

COPPER (29)

Copper is represented in the solar spectrum by the ultimate doublet 3273, 3247 (1^2S-m^2P), which is strong. The pair 5700, 5782 ($x-1^2P$) is probably also present. The former lines are too far in the ultra-violet to have been studied in the stars, and the latter are too faint.

ZINC (30)

The principal singlet $1S-1P$ is at 2138, and has therefore not been observed in stellar spectra. The 1^3P-1^3S lines at 4722, 4810, are seen in the stellar sequence, where they appear at F0, and have a maximum [202] at G0.

---

[200] Walters, J. Wash. Ac. Sci., **14**, 408, 1924.
[201] Russell, Ap. J., **55**, 119, 1922.
[202] Menzel, H. C. 258, 1924.

RUBIDIUM, STRONTIUM, YTTRIUM 81

Two unclassified lines of ionized zinc are mentioned in Fowler's Report as lying at 5894, 6214. Neither of these lines can be traced in solar or stellar spectra.

GALLIUM (31)

The occurrence of gallium in stellar spectra is confined to the identification of two solar lines by Hartley and Ramage.[203] The lines in question are at 4033, 4172, and are the ultimate lines of the element ($1^2P - m^2S$). They are too faint to be studied in the stars.

RUBIDIUM (37)

The ultimate lines of rubidium have been detected in the sunspot spectrum,[204] but they are not found in the spectra of the sun or stars.

STRONTIUM (38)

The element strontium is of great astrophysical importance, owing to the use of its enhanced lines in the estimation of absolute magnitudes. The neutral atom is represented in the sun and stars by the ultimate line ($1S - mP$) 4607, which is first clearly seen[205] at F0, and increases progressively in strength with advancing type. It varies with absolute magnitude, being weakened in stars of high luminosity later than K0. Estimates for the intensity of this line are difficult with small dispersions, as it is blended in cool stars.

IONIZED STRONTIUM

Ionized strontium is represented in stellar spectra by the $1^2S - m^2P$ and the $1^2P - m^2S$ series. The former contains the important absolute magnitude lines 4215, 4077, which are first seen at about A0, and reach a maximum[206] near K2. They appear to have "abnormal" intensities in certain stars,[207] and in the A stars are often the finest and sharpest lines in the spectrum. This behavior suggests a high-level origin, but " stationary

[203] Ap. J., **9**, 214, 1899. [204] Russell, Ap. J., **55**, 119, 1922.
[205] Menzel, H. C. 258, 1924. [206] Chapter VIII, p. 126.
[207] Chapter XII, p. 169.

strontium," although suggested by Plaskett [208] as likely to occur, has not yet been observed.

YTTRIUM (39)

Numerous lines of yttrium [209] are found in the solar spectrum. The lines of the ionized atom are somewhat stronger than the lines of the neutral atom. The lines of the neutral element which can be identified in the solar spectrum are contained in the following table.

TABLE XV

Series	Wave-Length	Int.	Attribution	Int.	Wave-Length
$^2D_3-^2P_2$	3620.94	20	Y?	∞	3621.110
$^2D_2-^2P_1$	3592.91	10	Y	0	3593.040
$^2D_2-^2P_2$	3552.69	3		–	–
$^2D_3-^2P_2$	4128.32	30		∞	4128.46
$^2D_2-^2P_2$	4039.83	5	Y	∞	4040.013
$^2D_2-^2P_1$	4047.65	8	Y	0N	4047.823
$^2D_3-^2F_4$	4102.38	20	Y	0	4102.541
$^2D_3-^2F_3$	4167.52	10		∞	4167.737
$^2D_2-^2F_3$	4077.39	20	La, Y	1Nd?	4077.498

The multiplets $^2D-^2P$ at 4174, etc., and $^2D-^2F$ at 4674, etc., and the $^2D-^2D'$ multiplets, do not appear in the solar spectrum. None of the above lines is strong enough to be seen in the spectra of the stars.

IONIZED YTTRIUM

Four of the multiplets attributed to ionized yttrium [210] are satisfactorily identified in the solar spectrum. The wave-lengths and identifications are contained in Table XVI, p. 83. The arrangement is as in Table XI.

[208] Pub. Dom. Ap. Obs., 2, 287, 1924.
[209] Meggers, J. Wash. Ac. Sci., 14, 419, 1924.
[210] Ibid.

IONIZED YTTRIUM

TABLE XVI

Series	Wave-Length	Int.	Attribution	Int.	Wave-Length
*$^3D_3-^3F_4$	3710.30	100	Y	3	3710.431
$^3D_3-^3F_3$	3832.87	20		3N	3833.026
*$^3D_2-^3F_3$	3774.33	50	Y	3	3774.473
$^3D_3-^3F_2$	3878.27	4		1	3878.334
$^3D_2-^3F_2$	3818.37	10	Y	1	3818.487
$^3D_1-^3F_2$	3788.69	30		2	3788.839
$^3D_3-^3D_3$	3600.72	50	Y	3	3600.880
$^3D_2-^3D_3$	3548.99	20	Y?	2	3549.151
$^3D_3-^3D_2$	3664.59	20	Y	2	3664.760
$^3D_2-^3D_2$	3611.05	30	Y, Mg?	2	3611.189
$^3D_1-^3D_2$	3584.51	10	Y	2	3584.660
$^3D_2-^3D_1$	3628.70	10	Y, Mg?	2	3628.847
$^3D_1-^3D_1$	3601.91	20	Y	1	3602.060
$^3D_3-^3P_2$	4309.61	20		1	4309.792
$^3D_2-^3P_2$	4235.71	6		0	4235.894
$^3D_1-^3P_2$	4199.283	3		00	4199 434
$^3D_2-^3P_1$	4398.03	15	In zircon*	1	4398.178
$^3D_1-^3P_1$	4358.72	8	Y–Zr	0	4358.879
$^3D_1-^3P_0$	4422.60	10	Fe, Y	3	4422.741
$^3F_4-^3F'_4$	5087.42	10	Y?	1	5087.601
$^3F_3-^3F'_4$	4982.12	3		000	4982.319
$^3F_4-^3F'_3$	5320.77	1		–	—
$^3F_3-^3F'_3$	5205.71	10	Y	0	5205.897
$^3F_2-^3F'_3$	5119.10	3		00	5119.292
$^3F_3-^3F'_2$	5289.81	2		000	5289.988
$^3F_2-^3F'_2$	5200.41	8	V	0	5200.590

* But not Zr.

ZIRCONIUM (40)

The ultimate lines of the zirconium atom [211] are all found in the solar spectrum, far into the ultra-violet.

The bands in the spectra of stars of Class S are found to correspond with those of ZrO_2, zirconium oxide.[212] A comparison

[211] de Gramont, C. R., **171**, 1106, 1920.
[212] Merrill, Pub. A. S. P., **33**, 206, 1921.

of the furnace spectrum of zirconium oxide with that of titanium oxide, which produces the characteristic flutings in Class M, indicates that titanium oxide persists to lower temperatures.[213] It is of interest to note that the only oxides, other than water, which have been detected in stellar spectra, are those of elements in the fourth column of the periodic table, namely carbon, titanium and zirconium. Probably this has a chemical interpretation. The presence of silicon dioxide SiO_2 has not been detected, although it might be anticipated.

NIOBIUM (41)

Rowland identifies some of the lines associated with niobium in the solar spectrum. The series relations are unknown, and the lines are too faint to be detected in stellar spectra.

MOLYBDENUM (42)

All the ultimate lines of molybdenum are present in the solar spectrum. They are too faint to be detected in the stars. The spectrum has been analyzed into series by Kiess,[214] and by Catalan.[215]

RUTHENIUM (44) RHODIUM (45) PALLADIUM (46)

The strongest lines in the spectra of the three lighter platinum metals are all present in the solar spectrum,[216] but are too faint to be traced in the spectra of stars. Series relations are as yet unknown. The heavier platinum metals, osmium (76), iridium (77), and platinum (78), are not certainly found in the solar spectrum.

SILVER (47)

The ultimate ($1^2S - m^2P$) lines of silver are at 3281 and 3383. They are both faintly present in the solar spectrum. The doublet at 4669, 4476 ($1^2P - m^2S$) is also probably present in the spectrum of the sun. The lines cannot be traced in stellar spectra.

[213] King, Pub. A. S. P., 36, 140, 1924.
[214] Kiess, Bur. Stan. Sci. Pap. 474, 1924.
[215] Catalan, An. Soc. Espan. Fis. y Quim., 21, 84 and 213, 1923.
[216] Russell, Science, 39, 791, 1914.

TIN, BARIUM, RARE EARTHS 85

TIN (50)

A line of neutral tin was reported by Lunt [217] in the spectrum of α Scorpii. Series relations are as yet unknown. In view of the absence of identifications of related lines, the attribution cannot be regarded as very certain. The strongest line in the spectrum of ionized tin is stated by the same investigator to lie at 4585.80. This line is either absent from or exceedingly weak in the solar spectrum.

BARIUM (56)

Neutral barium is not certainly present in stellar spectra. The first two ultimate lines ($1S-mP$) fall in the red and the ultra-violet respectively, and would therefore escape notice in the stars. No lines of corresponding wave-length appear in Rowland's table, but they are found in sunspots.[218] The $P-P'$ groups are also absent from the solar spectrum.

The strength of the barium lines in the sun has been thought by Russell to be abnormal, and the question has been considered by several investigators.[219–221]

IONIZED BARIUM

Ionized barium is represented by the 1^2P-m^2D lines which are present, though weak, in the sun, and by the 1^2S-m^2P lines, which appear [222] at A3, and increase in intensity for all cooler stars. The 1^2S-m^2P line is at 4555, and its behavior is difficult to trace, as it is much blended.

THE RARE EARTHS (57–71)

The spectra of the rare earths are so rich in lines that spurious identifications with lines in stellar spectra are often likely to occur. Numerous attributions to lanthanum (57), cerium (58),

---

[217] M. N. R. A. S., **77**, 487, 1907.
[218] Russell, Ap. J., **55**, 119, 1922.
[219] Saha, Phil. Mag., **40**, 472, 1920.
[220] H. H. Plaskett, Pub. Dom. Ap. Obs., **1**, 325, 1922.
[221] M. C. Johnson, M. N. R. A. S., **84**, 516, 1924.
[222] Menzel, H. C. 258, 1924.

86 ELEMENTS IN STELLAR ATMOSPHERES

and neodymium (60) are given in Rowland's table. The occurrence of some of these elements has also been definitely established in the spectra of certain stars. Kiess [223] has demonstrated the presence of europium (63) and of terbium (65) in α Canum Venaticorum. Numerous lines, identified with those of rare earth elements, are reported by Mitchell [224] in the chromospheric spectrum.

If the lines in the chromosphere and the A star are indeed derived from the rare earths, the atoms concerned must be at least singly ionized. No series relations have as yet been published for any rare earth element, excepting a short list of relative term values for lanthanum.[225] From analogy with the previous long period it would seem unlikely that the first ionization potential of these atoms can be as great as 13 volts, the value which would be required if the lines have a maximum intensity at A0.

LEAD (82)

A single line is attributed to lead in Rowland's table.

RADIUM (88)

Giebeler and Küstner [226] suggested the occurrence of radium lines in the chromosphere. The identification was discussed by Dyson [227] and by Mitchell.[228] In the light of later knowledge it appears improbable that an element so heavy, and terrestrially so rare,[229] would be present in the sun at sufficient heights and in great enough quantity to appear in the flash spectrum. The identification is probably to be regarded as spurious.

ELEMENTS NOT DETECTED IN STELLAR SPECTRA

The table which follows contains a list of elements which are absent, or of doubtful occurrence. The rare earths are omitted from the list.

[223] Kiess, Pub. Obs. Mich., 3, 106, 1923.
[224] Mitchell, Ap. J., 38, 407, 1913.
[225] Gousmid, Naturwiss., 41, 851, 1924.
[226] A. N., 191, 393, 1912.
[227] A. N., 192, 82, 1912. [228] A. N., 192, 266, 1912.
[229] Chapter I, p. 5.

ELEMENTS NOT DETECTED 87

The elements marked "doubtful" in the list are those for which coincidences with very faint solar lines occur.[230] Twelve out of the twenty-nine elements enumerated are halogens, inert gases, or metalloids, and it is significant that all the elements of these groups are absent, with the sole exception of helium.

At. No.	Element	Remark	At. No.	Element	Remark
4	Beryllium	Doubtful	53	Iodine	Absent
5	Boron	Absent	54	Xenon	Absent
9	Fluorine	Absent	73	Tantalum	Doubtful
10	Neon	Absent	74	Tungsten	Doubtful
15	Phosphorus	Absent	76	Osmium	Doubtful
17	Chlorine	Absent	77	Iridium	Doubtful
18	Argon	Absent	78	Platinum	Doubtful
32	Germanium	Doubtful	79	Gold	Absent
33	Arsenic	Absent	80	Mercury	Doubtful
34	Selenium	Absent	81	Thallium	Doubtful
35	Bromine	Absent	83	Bismuth	Doubtful
36	Krypton	Absent	86	Radon	Absent
49	Indium	Doubtful	90	Thorium	Doubtful
51	Antimony	Absent	92	Uranium	Doubtful
52	Tellurium	Absent			

[230] Abbot, The Sun, 92, 1911.

PART II
THEORY OF THERMAL IONIZATION

CHAPTER VI

THE HIGH-TEMPERATURE ABSORPTION SPECTRUM OF A GAS

IT is certain that the conditions of which we see the integrated result in the stellar spectrum are exceedingly complicated. Unfortunately, the superficial portion of the star about which direct observational evidence can be obtained is far less tractable to theory than is the interior. Progress is only made possible by treating at the outset a simplified case, by aiming merely at approximate results, and in particular by limiting the preliminary discussion to the factors which are numerically the most effective. As an introduction to the theory of thermal ionization, the present chapter aims at the reconstruction and interpretation of a stellar spectrum by applying known physical laws under very simple conditions.

The stellar reversing layer may be represented by an optically thin layer of gas, at a pressure of the order of one ten thousandth of an atmosphere; it lies between the observer and a photosphere which radiates as a black body. The observer receives the radiation from both reversing layer and photosphere, which are regarded, in the present descriptive section, as independent. A more complete treatment would take account of the temperature and pressure gradients in the atmosphere of the star, the flux of energy, and the consequent intimate connection between reversing layer and photosphere. Actually they grade imperceptibly one into the other. The photosphere is that level in the atmosphere at which the general opacity cuts off the direct light from the interior;[1] in the case discussed the reversing layer is considered to be optically so thin that the general opacity is negligible. The *selective* opacity, depending on the natural absorption frequencies of the atoms present in the gas, gives rise

[1] Stewart, Phys. Rev., 22, 324, 1923.

to the line absorption spectrum which we are about to consider; the region of sensible *general* opacity, represented by the photosphere, gives rise to a continuous spectrum corresponding to the continuous background in the star.

THE ABSORPTION OF RADIATION

The light passing through the layer of gas is absorbed, in terms of atomic theory, in the shifting of an electron from one energy level in an atom to some higher level, losing in the process energy of the definite frequency which is associated with that particular atom and energy transfer. The energy levels and possible electron transfers for the hydrogen atom are reproduced in Figures 2 and 3. In Figure 3 the horizontal lines represent the stationary states which can be assumed by the electron, and the arrows denote possible jumps from one stationary state to another. In Figure 2 the electron orbits corresponding to some of the simpler corresponding transitions for the hydrogen atom are represented. Arrows denote transfers from one orbit to another. The designation of the line corresponding to each transfer is appended to the appropriate arrow. It is evident that the occurrence of a given jump requires that there shall be an electron in the stationary state from which the jump originates.

The *ultimate lines*[2,3] are those which arise from the lowest energy level, and are therefore those most readily absorbed by the normal (undisturbed) atom. In the hydrogen spectrum these comprise the Lyman series,[4] with the first member at 1215.68. The Balmer and Paschen series are both *subordinate* series, requiring an initial lifting of the electron from the lowest energy level into a two and three (total) quantum orbit, respectively. The absorption of the Lyman line Ly α is necessary to a hydrogen atom before it is in a fit condition to absorb any Balmer line, and for the absorption of a Paschen line, an initial absorption of Ly β or H α is required.

[2] De Gramont, C. R., **171**, 1106, 1920.
[3] Russell, Pop. Ast., **32**, 620, 1924.
[4] A. Fowler, Report on Series in Line Spectra, 1922.

It appears plausible to assume, at least for low partial pressures, that the amount of energy of any frequency that is lost by black-body radiation in passing through the absorbing layer will vary jointly with the supply of energy and the number of atoms which are in a suitable state to absorb that particular frequency. One of the problems that arise is therefore that of determining what fraction of the whole number of atoms of a given kind will be able to absorb. It is to this problem that ionization theory is able to offer a solution.

By choosing the much simplified case of very low pressure and small concentration, the effects of ionization by collision [5] and of nuclear fields are probably eliminated. The remaining factor which may influence the number of absorbing atoms is thermal ionization, and this is actually the numerically important factor, as was first pointed out by Saha.[6] It is of interest to note that Saha's original treatment contemplated pressures of the order of one atmosphere. Under such conditions the effects of collisions and of nuclear fields are not negligible, and might well have invalidated the theory. Later work has shown conclusively, however, that the pressures in the reversing layer are probably not greater than 10^{-4} atmospheres,[7,8] and that thermal ionization is the predominant factor under these conditions.

The absorbing layer is to be regarded as consisting of a mixture of all chemical elements, without any assumption as to quantity, so long as the partial pressure of each individual element is low. In other words, no account is taken, at the present stage, of the relative abundances of different kinds of atoms — the *total* effectiveness of the corresponding elements as absorbers. The *changes* in the absorption of the black body radiation by a given element with changing temperature will be the same whatever the partial pressure, provided it is low, and it is with these changes that the preliminary schematic discussion is concerned.

[5] R. H. Fowler, Phil. Mag., **47**, 257, 1924.
[6] Proc. Roy. Soc., **99A**, 135, 1921.
[7] M. N. R. A. S., **83**, 403, 1923; *ibid.*, **84**, 499, 1924.
[8] Russell and Stewart, Ap. J., **59**, 197, 1924.

LOW TEMPERATURE CONDITIONS

At low temperatures all the elements will tend to be in their normal atomic state, unless they are aggregated into molecules or compounds. At temperatures of 2500°, which is about the lower limit encountered in dealing with stellar spectra, there is evidence of the existence of various oxides (CO, TiO_2, ZrO_2), of "cyanogen," and of hydrocarbons, but most of the other possible compounds appear either to be dissociated or to be in very low concentration. Probably the normally polyatomic gases such as oxygen, nitrogen, and sulphur, are to some extent present in the molecular state. Even at atmospheric pressure all the metals are vaporized at 2500° excepting tantalum and the platinum metals, which boil at about 2800° under a pressure of 760 mm; at lower pressures the temperature of vaporization is, of course, lower. The metallic molecule appears normally to be monatomic, so that it will give its line spectrum unless it is in combination. The fact that silicon, the most refractory substance, excepting carbon, with which we have to deal, gives its line spectrum in the coolest stars known, indicates that all the elements may be considered to be gaseous in stellar atmospheres.

ULTIMATE LINES

The absorption spectrum given by the reversing layer when it is at a low temperature will consist of the lines given most readily by the atom in its normal state. The energy transfers which move an electron from its normal orbit to another correspond to the "ultimate lines," and these lines will therefore be especially outstanding in the spectra of the coolest atmospheres. They are of such importance, from theoretical and from practical standpoints, that a list of them is reproduced here. Successive columns of the table give the atomic number and atom, the ionization potential, the wave-lengths of the ultimate lines, and an indication of their observed occurrence in stellar spectra. An asterisk denotes that the line has been observed, and a dash indicates that it has not been recorded.

ULTIMATE LINES

It may be remarked that the ultimate lines of sodium, potassium, lithium, rubidium, and caesium are in the visible region — a fact which is utilized in the laboratory flame tests used in qualitative analysis.[9] The brilliancy of the flame colors obtained in the Bunsen burner, at the temperature of about 1500°C., is a striking elementary illustration of the readiness with which the atom in its normal state will take up and re-emit the frequency corresponding to the ultimate lines (second pair for K, Rb, Cs).

TABLE XVII

Atom		Ionization Potential	Wave-length	Stellar Occurrence
1	H	13.54	1215	—
2	He	24.47	584, 557	—
3	Li	5.37	6707	*
4	Be	?	2349	—
5	B	?	2498, 2497	—
6	C	?	2479	—
7	N	?	—	—
8	O	13.56	1306, 1304, 1302	—
9	F	?	—	—
10	Ne	16.7	—	—
11	Na	5.12	5896, 5890	*
12	Mg	7.61	2852	—
13	Al	5.96	3962, 3944	*
14	Si	?	2882	—
15	P	?	2553, 2536	—
16	S	10.31	1915, 1900	—
17	Cl	?	—	—
18	A	?	—	—
19	K	4.32	7699, 7665	*
20	Ca	6.09	4227	*
21	Sc	?	4247, 3652	*
22	Ti	6.5	5065, 5040, 5014	*
23	V	?	4331, 4333	*
24	Cr	6.75	4290, 4275, 5254	*
25	Mn	7.41	4034, 4033, 4031	*
26	Fe	?	2756, 2749	—
27	Co	?	3454, 3405	*
28	Ni	?	3415	*

[9] Eder and Valenta, Atlas Typischer Spektren, 10, 1911.

96   ABSORPTION SPECTRUM OF A GAS

Atom	Ionization Potential	Wave-length	Stellar Occurrence
29 Cu	7.69	3274, 3248	*
30 Zn	9.35	2139	–
31 Ga	5.97	4172, 4033	*
37 Rb	4.16	7948, 7800	*
38 Sr	5.67	4607	*
40 Zr	?	4496, 4392	*
42 Mo	?	3903, 3864, 3798	*
47 Ag	7.54	3383, 3281	*
48 Cd	8.95	2288	–
49 In	5.76	4511, 4102	*
50 Sn	?	3262	?
55 Cs	3.88	8943, 8581	–
56 Ba	5.19	7911	*
57 La	?	3949	*
79 Au	8.72	2676, 2428	–
80 Hg	10.39	2537, 1850	–
81 Tl	6.08	5350, 3775	–
82 Pb	7.38	4058, 3684	*

It is possible to predict from the table which lines are likely to appear in the spectra of the coolest stars. The ultimate lines of Al, K, Ca, Sc, Ti, V, Cr, Mn, Co, Ni, Cu, Ga, Rb, Sr, Zr, Mo, Ag, In, Ba, La, and Pb fall in the region ordinarily photographed, and Na can be reached in the yellow. All these elements in the neutral state would therefore be anticipated in the spectra of cooler stars, and they are indeed found without exception. The ultimate lines of several elements contained in the list lie in the far ultra-violet, and cannot be detected in stellar spectra. The corresponding neutral elements will therefore not be recorded unless they also give a strong subordinate series in the photographic region. The elements C, O, S, and probably N, all have ultimate lines in the ultra-violet, and possess no subordinate series in the appropriate range of wave-length. Their apparent absence in the neutral state from stellar spectra is therefore fully explained. All of these elements appear in the hotter stars in the once or twice ionized condition. The elements H, Mg, and Si have strong subordinate series in the photographic region — the Balmer series, the "b" group, and the line [10] at 3905, respectively. They are accordingly represented in the cooler stars.

The elements contained in the table and not yet discussed are Be, B, Ne, A, F, and Cl. These elements have not been detected in stellar spectra. In seeking an explanation of their apparent absence, it has been suggested that a low relative abundance of the corresponding atoms may be responsible. Arguments from terrestrial analogy must be applied with caution, but there is reason to suspect that they may here have a legitimate application.[11] It may be suggested that boron, beryllium, neon, and argon are present in the stars in quantities too small to be detected. The halogens are unrepresented, but it is not possible to draw useful inferences until their laboratory spectra are more fully analyzed.

IONIZATION

At the lowest temperatures, then, the ultimate lines will predominate. As the temperature of the absorbing layer is raised, ionization — the complete ejection of the electron from the atom, instead of a displacement from one stationary state to another — will set in, and the tracing of the resulting spectral changes is the salient feature of the Saha theory. "Ionization can be effected in many ways. To expel an electron against the attractive force of the remainder of the molecule, work is required, and the necessary energy may be furnished by X rays or γ rays, or by collision with other electrons. . . . At high temperatures, when the conditions of maximum entropy demands an appreciable amount of ionic dissociation, the requisite energy is drawn from the environment. . . . The work required to ionize a single molecule, when expressed as the number of volts through which an electron must fall to acquire this energy, is the *ionization potential*; it may be regarded as the latent heat of evaporation of the electron from the molecule" (Milne).[12]

[10] See Chapter V, p. 69. A. Fowler, Bakerian Lecture, 1924, designates this an ultimate line.
[11] See Chapter XIII, p. 185.   [12] Milne, Observatory, 44, 264, 1921.

The analogy between ionization and evaporation illustrates very well the scope of the Saha theory, in which the process is treated as a type of chemical dissociation. Corresponding to each temperature there is a definite state of equilibrium, where the forward and backward velocities of the ionization process are equal — in other words where ionization and recombination are proceeding at the same rate. The method of statistical mechanics has been applied to this problem by Fowler and Milne.[13] Here the analysis will not be reproduced, but the formulae are required in order to illustrate the process of ionization.

The number of atoms which are un-ionized at any given temperature is given by the expression

$$1 - x = \frac{b(T)}{b(T) + aT^{\frac{5}{2}}e^{-\chi_1/kT}}$$

where x = number of atoms ionized.
$b(T)$ = the "partition function."
$a = 0.332/P_e$, where P_e is the partial pressure of electrons.
T = absolute temperature.
k = Boltzmann's constant, $= 1.37 \times 10^{-16}$.
χ_1 = the ionization potential.

This is the number of atoms which is effective in absorbing the ultimate lines at that temperature. For low values of T, the number of un-ionized atoms falls off at first very slowly with rising temperature, up to a point depending only on the ionization potential. Beyond this temperature the number of neutral atoms falls off with great rapidity. The diagram (Figure 5) illustrates the fall in the number of neutral atoms, and the consequent decay in strength of the ultimate lines. So steep is the gradient of $1-x$ at the higher temperatures that the quantity is best plotted logarithmically. The ultimate lines will persist, with almost undiminished intensity, up to the temperature

[13] M. N. R. A. S., **83**, 403, 1923; **84**, 499, 1924.

IONIZATION 99

at which the gradient of $1-x$ begins to increase. This critical temperature increases with ionization potential, and neutral

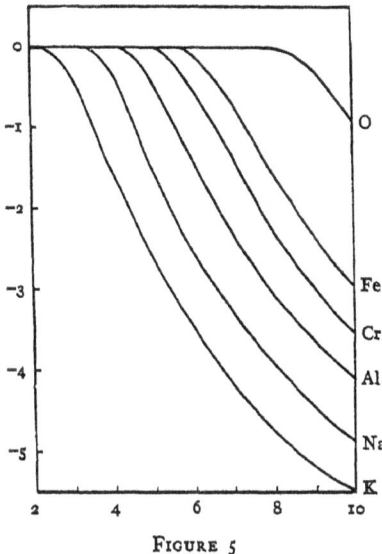

FIGURE 5

Ultimate lines of neutral atoms. Ordinates are logarithms of computed fractional concentrations; abscissae are temperatures in thousands of degrees. The curves show the decrease in the number of neutral atoms, with rising temperature, and the consequent decay in strength of the ultimate lines, for the atoms indicated on the right margin.

atoms of high ionization potential should display very persistent ultimate lines as the temperature rises.

As ionization becomes more and more complete, the intensity of the ultimate lines falls off until so small a number of neutral atoms remains that their lines cease to appear in the absorption spectrum.

SUBORDINATE LINES

The neutral atom gives rise to other lines besides the ultimate lines, but these require the transfer of an electron from some stationary state, not the normal one, to another stationary state. The atom must receive a definite quantity of energy, equal to the excitation potential of the initial stationary state, in order to be in a condition to absorb a line of a subordinate

ABSORPTION SPECTRUM OF A GAS

series which originates from that state. If there is an appreciable energy supply, a certain fraction of the neutral atoms pres-

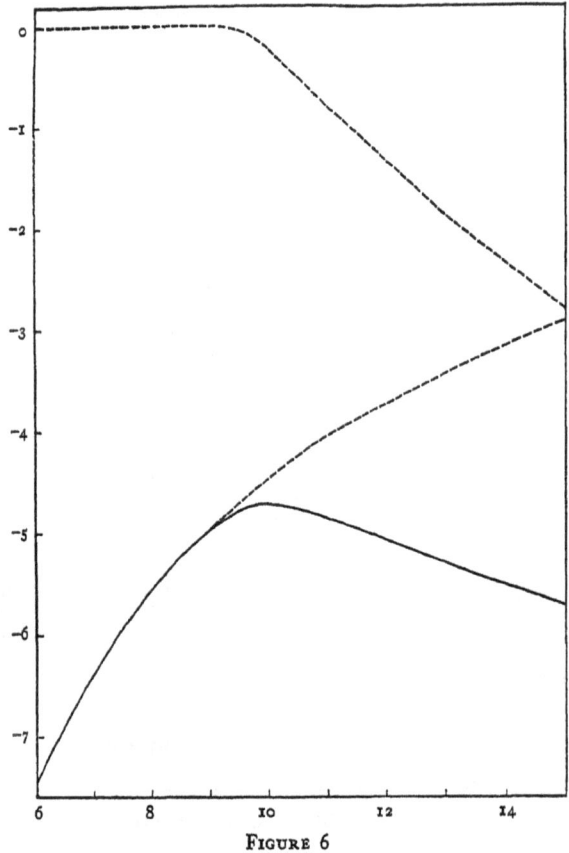

FIGURE 6

Production of the maximum of an absorption line. Ordinates are logarithms of computed fractional concentrations; abscissae are temperatures in thousands of degrees. The curves reproduced are those for the Mg + line at 4481. The upper broken curve represents the fraction of magnesium atoms that is singly ionized at the corresponding temperature; the lower broken curve represents the fraction of the Mg + atoms present that is in a suitable state for the absorption of 4481. The full line represents the sum of the ordinates of the dotted curves, and gives the fraction of the total number of magnesium atoms that is able to absorb 4481 at the various temperatures indicated by the abscissae.

ent will have received this excitation energy, which is of course smaller than the ionization potential, and these atoms will be in a position to absorb the subordinate series. The fraction, f_r, of the total number of neutral atoms which have become able to absorb the lines associated with a definite excitation potential is given by Fowler and Milne as

$$f_r = \frac{q_r e^{-(\chi_1 - \chi_1^{(r)})/kT}}{b(T)}$$

where $(\chi_1 - \chi_1^{(r)}) =$ excitation potential. The quantity f_r increases with the temperature, approaching the value unity asymptotically.

The total number of atoms active in absorbing a subordinate series at any temperature is evidently the product of the number of *neutral* atoms and the quantity f_r. The curves for these two quantities are plotted logarithmically in Figure 6, the magnesium line 4481 being used as an illustration. The total number of absorbing atoms may be obtained by adding the ordinates. It will be seen that the number of such atoms increases, passes through a maximum and decreases again, as the temperature is raised. The maximum for a surbordinate line of the neutral atom may occur, as in the case of helium, when ionization is far advanced.

In the special case where $(\chi_1 - \chi_1^{(r)}) = 0$, the second curve, which represents the growth of the fraction f_r, becomes a straight line parallel to the temperature axis, and the first, or ionization, curve, approaches the zero ordinate asymptotically at low temperatures. The ordinate of the curve representing f_r is zero, and the resultant sum gives a curve identical with the curve for the ultimate lines. Ultimate lines thus appear as the special case of subordinate lines for which the excitation potential is zero. This fits exactly with the definition of ultimate lines as the lines naturally absorbed by the cold vapor — no initial excitation is required to bring the atoms into a state in which they can absorb.

LINES OF IONIZED ATOMS

As soon as ionization sets in, the absorbing layer begins to contain a new kind of atom, derived from the neutral atoms by the complete ejection of one electron. These ionized atoms will absorb their own spectrum, which differs completely from that of the corresponding neutral atom; and the degree of absorption will again depend on the number of such ionized atoms present in the reversing layer.

The ionized atom has in general a spectrum corresponding exactly to that of the neutral atom preceding it in the periodic table, but with a different Rydberg constant.[14,15] Two types of lines arise, as before — ultimate and subordinate lines. For the number of atoms which can absorb the ultimate lines of the enhanced spectrum, the formula reduces to

$$n_r = \frac{q_r e^{-(x_1-x_1^{(r)})/kT}}{b(T)+\dfrac{P_e}{0.332\,\sigma}T^{\frac{5}{2}}e^{-(x_1/kT)}+\dfrac{0.332\,\sigma'}{P_e}T^{-\frac{5}{2}}e^{x_2/kT}}$$

Account is here taken of the residual neutral atoms by the middle term of the denominator, which is very small, and is

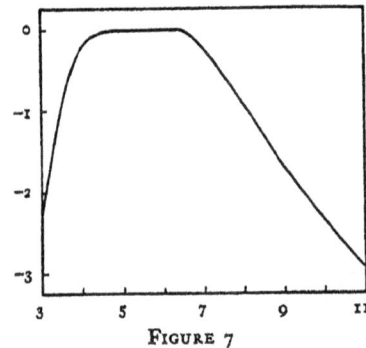

FIGURE 7

Maximum of the ultimate line of an ionized atom. Ordinates are logarithms of computed fractional concentrations; abscissae are temperatures in thousands of degrees. The curve is drawn for the line 4554 of Ba+, on the assumption that Pe is 1.3×10^{-4} atmospheres.

only of sensible magnitude for the ultimate lines, when the numerator is equal to unity. The following curve shows the number of absorbing atoms. The flatness of the maximum is especially to be noted, suggesting that the ultimate lines of the

---

[14] Sommerfeld, Atombau und Spektrallinien, 457, 1922.
[15] Meggers, Kiess, and Walters, Journ. Op. Soc. Am., **9**, 355, 1924.

ionized atom, like the ultimate lines of the neutral atom, will be very persistent. The H and K lines of Ca+, and the corresponding lines 4077 and 4215 of Sr+, and 4555 of Ba+, would thus be expected to show over a considerable range in temperature and spectrum, and this is actually found to be the case.

The subordinate lines behave substantially as do the subordinate lines of a neutral atom, rising to a maximum at a temperature which depends chiefly on the ionization potential. It is assumed in deriving the corresponding equations that in practice the number of surviving neutral atoms will be too small to affect the concentration of ionized atoms giving the subordinate lines. This assumption may be shown to be justified at maximum intensity of the absorption line, though possibly the neutral atoms are not always negligible at the first appearance of the ionized lines of a very abundant atom.

SUMMARY

The general results of raising the temperature of the absorbing layer have now been traced. Although a greatly simplified case has been considered, the observed changes in the stellar spectral sequence have been very satisfactorily predicted.

At low temperatures the lines of neutral atoms are strong, in particular the ultimate lines, such as 3930 of Fe, 3999 of Ti, 4254 of Cr, and 4033 of Mn, which are at maximum strength, and decrease at first slowly, then rapidly in the hotter stars. The subordinate lines of neutral atoms, 4455 of Ca and 4352 of Fe, for example, attain a maximum, and then fall off with rising temperature. For many of the metallic lines for which no maximum is recorded, like those of the subordinate series of Na, the theoretical maximum is at a temperature equal to that of the coolest stars examined. Atoms with ionization potential less than 5 volts will in general give maxima below 3000°.

As the temperature increases, the lines of ionized atoms begin to appear, the ultimate lines rising very quickly in intensity, and persisting almost at maximum over several spectral classes. Later in the sequence the subordinate series for ionized atoms appear, rise to a sharper maximum, and fade more rapidly. The 4481 line of Mg+, the 4267 line of C+, and the 4128 line of Si+, show this effect well.

As the fall of intensity of the lines of neutral atoms after maximum is the result of the progress of ionization, it would be expected that the lines of the ionized atom would appear while those of the neutral atom were still quite strong, and that the one series would rise in strength as the other decreased. The lines of the neutral atom may persist over a large part of the range of the ionized lines. This is the case with the 4227 line of Ca, which persists until Class A0, while the H and K lines of Ca+ have been visible throughout the whole spectral sequence, and have been decreasing in intensity from K0 onwards, owing chiefly to the rise of second ionization and the consequent formation of Ca++, which gives a spectrum in the ultra-violet and is therefore not detected in the stars.

As the temperature is further raised, the second and third ionizations set in, and presumably follow the same procedure as has been outlined for less ionized atoms. The lines of N++, C++, Si++, and Si+++ will serve as examples. The lines of the doubly ionized atoms of the metals are in general in the ultra-violet portion of the spectrum, and the corresponding elements do not therefore appear in the hotter stars, where they would otherwise be anticipated.

Qualitatively the prediction of the theory of ionization is fully satisfied. The quantitative discussion involves more rigorous treatment, and is reserved for a later chapter.

CHAPTER VII

CRITICAL DISCUSSION OF IONIZATION THEORY

THE theory of thermal ionization, of which the preceding chapter contains an illustrative discussion, may be treated from two points of view — the sufficiency of the analytical treatment, and the nature of the underlying physical assumptions. Actually the two questions are merely two different ways of regarding the validity of the theory, but they divide the discussion conveniently into a section dealing with the analytical treatment and a section dealing with the physical assumptions.

The original treatment by Saha[1] was based on the Law of Mass Action, and the application to stellar atmospheres raised questions of a physical rather than of an analytical nature. These questions are fundamental not only to the Saha treatment, but also to the more recent development of the theory, and they will be discussed in the second half of the present chapter. The first half will be devoted to the analytical formulae.

MARGINAL APPEARANCE AND MAXIMUM

Saha's discussion was based on the observation of "marginal appearance" — the spectral class at which a particular absorption line is at the limit of visibility. The use of this quantity as a criterion for the temperature scale has certain practical drawbacks. Marginal appearance depends directly on relative abundance, since a more abundant element will give visible lines at a lower "fractional concentration," that is to say, when a smaller fraction of the element is contributing to the lines in question. Further, in estimating the intensities of lines in stellar spectra, difficulty is experienced when the lines are faint, and the spectral class at which they are first or last seen depends on

[1] Proc. Roy. Soc., **99A**, 136, 1921.

their width and definition, the intensity of the continuous background, the presence of other lines, and the dispersion used. All of these factors are subject to variation, and in particular the intensity distribution in the continuous background changes with the temperature. The statistical theory of Fowler and Milne has, therefore, a great advantage in that it leads to an estimate of the temperature at which a given line attains maximum. A maximum, unlike a marginal appearance, can be determined without ambiguity from homogeneous material, whatever the dispersion. In the cooler stars the estimates may be made difficult by blending, but the uncertainty can generally be removed by examining the maxima of several related lines. Of the observational factors enumerated above as affecting the estimation of marginal appearance, the changing intensity distribution of the continuous background with the temperature is the only one that may prove serious for the method of maxima.

THEORETICAL FORMULAE

The theory developed by Fowler and Milne has been exhaustively discussed by these authors in several papers, and it appears unnecessary to reproduce the analysis in detail. The ionization and excitation curves have been treated diagrammatically in the previous chapter. The detailed formulae follow.

"If χ_1 is the ionization potential of the atom, $\chi_1^{(r)}$ the (negative) energy of a given excited state, then for a given partial electron pressure P_e of free electrons, the temperature T of maximum concentration of atoms in the given excited state $\chi_1^{(r)}$ is given by [2]

$$P_e = \frac{\chi_1^{(r)} + \tfrac{5}{2}kT}{\chi_1 - \chi_1^{(r)}} \cdot \frac{(2\pi m)^{\tfrac{3}{2}} (kT)^{\tfrac{5}{2}} \sigma_1}{h^3 b_1(T)} \cdot e^{-\chi_1/kT}$$

$m =$ mass of electron.
$k =$ Boltzmann's constant.
$h =$ Planck's constant.

[2] M. N. R. A. S., **83**, 403, 1923; *ibid.*, **84**, 499, 1924.

ANALYTICAL FORMULAE 107

σ = symmetry number of the atom (number of spectroscopic valency electrons).
$b(T)$ = the partition function
$= q_1 + q_1^{(2)} e^{-(x_1 - x^{(2)})/kT} + q_1^{(3)} e^{-(x_1 - x^{(3)})/kT} + \ldots$
q = weight of corresponding stationary state.

" The temperature at which the concentration of singly ionized atoms reaches a maximum is given by [3]

$$P_e = \left[\frac{\sigma_1 \sigma_2}{b_1(T) b_2(T)}\right]^{\frac{1}{2}} \left[\frac{\chi_2 + \frac{5}{2} kT}{\chi_1 + \frac{5}{2} kT}\right]^{\frac{1}{2}} \frac{(2\pi m)^{\frac{3}{2}} (kT)^{\frac{5}{2}}}{h^3} e^{-\frac{1}{2}(\chi_1 + \chi_2)/kT}$$

χ_2 = second stage ionization potential.
σ_2 and $b_2(T)$ refer to the singly ionized atom.

The two formulae that have been quoted assume, in effect, that at any stage of ionization the number of atoms, in stages other than those whose constants appear in the formulae, is negligible. When two ionizations occur in very close succession, this assumption no longer holds, and the equations, as modified to embody the necessary correction, are as follows.[4]

For the subordinate series of a neutral atom,

$$P_e = \frac{\chi_1^{(r)} + \frac{5}{2} kT}{\chi_1 - \chi_1^{(r)}} \cdot \frac{(2\pi m)^{\frac{3}{2}} (kT)^{\frac{5}{2}} \sigma_1}{b_1(T) h^3} e^{-\chi_1/kT} + \frac{1}{P_e} \frac{\chi_2 + \chi_1^{(r)} + 5 kT}{\chi_1 - \chi_1^{(r)}} \cdot$$
$$\frac{(2\pi m)^3 (kT)^5 \sigma_1 \sigma_2}{b_1(T) b_2(T) h^6} \times e^{-(\chi_1 + \chi_2)/kT}$$

" This equation must be used to calculate P_e whenever the ionization potential of the stage in question is closely *followed* by the ionization potential of the succeeding stage."

For the subordinate series of the ionized atom,

$$P_e = \frac{\chi_2^{(r)} + \frac{5}{2} kT}{\chi_2 - \chi_2^{(r)}} \cdot \frac{(2\pi m)^{\frac{3}{2}} (kT)^{\frac{5}{2}} \sigma_2}{b_2(T) h^3} - \frac{P_e^2 (\chi_1 + \chi_2 - \chi_2^{(r)}) + \frac{5}{2} kT}{\chi_2 - \chi_2^{(r)}} \cdot$$
$$\frac{b_1(T) h^3}{(2\pi m)^{\frac{3}{2}} (kT)^{\frac{5}{2}} \sigma_1} e^{\chi_1/kT}$$

[3] M. N. R. A. S., **83,** 403, 1923.
[4] M. N. R. A. S., **84,** 499, 1924.

This equation " must be used wherever the ionization potential of the stage in question is closely *preceded* by the ionization potential of the preceding stage. The corrections . . . result in making the maxima for the lines of the two stages occur farther apart in the temperature scale. If we express the correction in the form of a factor $(1+a)$ . . . then a is of the order $e^{-(\chi_2-\chi_1)/kT_{max}}$. Since kT_{max} varies roughly as χ_1 or χ_2, we see that the importance of the correction is determined by the closeness of χ_1/χ_2 to 1."

The values of n_r, the fractional concentration of the atom in question, are obtained through the application of the ordinary methods of statistical mechanics to the equilibrium between atoms and electrons in the reversing layer. The values of P_e at the maximum are obtained by differentiating the expression for n_r with respect to T, and equating to zero, since the maximum of absorption will occur when n_r is at a maximum.

The analytical treatment calls for no comment. Its basis has been fully discussed by R. H. Fowler [5] in a series of papers. The weights (q) of the atomic states employed were based on the work of Bohr [6] on the relative values of the a priori probabilities of the different stationary states for hydrogen. On this view, $b(T) = 1$ for all atoms excepting those of H and He+, for which it is equal to 2. The convergence of the series for $b(T)$ was not established by Fowler and Milne, but the authors regard the subsequent investigation by Urey [7] as justifying their assumption that " for physical reasons one must suppose the series effectively cut off after a certain number of terms. Usually the series then reduces (as regards its numerical value) practically to its first term."

PHYSICAL CONSTANTS REQUIRED IN THE FORMULAE

The application of the equations will of course depend upon an accurate knowledge of the constants involved. The quantities m, k, h, and q require no comment. The symmetry

[5] R. H. Fowler, Phil. Mag., **45**, 1, 1923.
[6] Bohr, Mem. Ac. Roy. Den., **4**, 2, 76, 1922.   [7] Ap. J., **59**, 1, 1924.

number σ_1 of the neutral atom is in effect the number of *spectroscopic* valency electrons given in Bohr's table,[8] for the atoms for which it is known. In all the applications made by Fowler and Milne the quantity was equated to 1 or 2, and it is very probable that this number is not in any case exceeded. For carbon, where the chemical valency is equal to 4, the value of σ_1 is still 2, as has been shown by Fowler's analysis of the spectrum of ionized carbon.[9] The value of σ_1 is not known for atoms in the long periods, but in the present work it is assumed to be 1 and 2 for atoms with arc spectra which show even and odd multiplicities, respectively. The uncertainty in the value of σ_1 introduces only a relatively small error into the result, since P_e depends on the first power of σ_1, and in no case considered can σ_1 exceed five.

The most important factor involved in the theory is P_e, the partial pressure of electrons in the reversing layer. By assuming P_e constant at about 10^{-4} atmospheres, and treating T as the unknown, a temperature scale which agrees substantially with those derived from measurements of radiation may be deduced from the observed positions of the maxima. The first discussion of the data then available was made by Fowler and Milne in their original paper.[10] Subsequent investigations of the positions of maxima have been published by Menzel[11] and by the writer.[12] These observations, and the scale derived from them, will be discussed in the two following chapters.

The value of P_e has been recently shown by several kinds of investigation to be at least as low as was assumed by Fowler and Milne, so that their assumption that a uniform mean pressure can be used, as a first approximation, in deriving a temperature scale from their formula appears to be justified. Milne[13] points out that " on whatever specific assumptions " the theory rests, " the mean pressure for a maximum of intensity in an absorption line is found to depend on the absolute value of the

[8] Chapter I, p. 9.
[9] Proc. Roy. Soc., 103A, 413, 1923.
[10] M. N. R. A. S., 83, 404, 1923.
[11] H. C. 258, 1924.
[12] H. C. 252, 256, 1924.
[13] Phil. Mag., 47, 209, 1924.

absorption coefficient. In fact . . . it is clear that the greater the absorbing power of the atoms in question, the more opaque is the stellar atmosphere in the frequency concerned, and so the greater the height and the smaller the pressure at which the line originates." That the absorption coefficient in the stellar atmosphere is very high is suggested by the reorganization times (" lives ") of such atoms as have been investigated,[14] and Milne's discussion of the life of the excited calcium atom from astrophysical data lends weight to the suggestion. A high absorption coefficient leads at once to low pressures in the reversing layer, and theory has gone far towards indicating that pressures of the order of 10^{-4} atmospheres are to be expected on a priori grounds.[15]

The observational evidence bearing on pressures in the reversing layer will be found [16] in Chapter III. The case appears to be a strong one, resting on evidence of many different kinds — notably pressure shifts, line sharpness, and series limits. Russell and Stewart,[17] in their exhaustive discussion of the question, conclude that "all lines of evidence agree with the conclusion that the total pressure of the *photospheric gases* is less than 0.01 atmosphere, and that the average pressure in the *reversing layer* is not greater than 0.0001 atmosphere."

The observational evidence gives the *total* pressure, but the partial electron pressure will not differ greatly from this. Although even in the hottest stars three ionizations is the greatest number observed, most of the elements that constitute the stellar atmosphere are appreciably ionized at temperatures greater than 4000°, so that the partial electron pressure is at least half the total pressure.

PHYSICAL ASSUMPTIONS

The method applied by Saha to stellar atmospheres was borrowed from physical chemistry. The Law of Mass Action, and the theory of ionization in solutions which is based upon it, have in general been very well satisfied in dilute solution.[18] The

[14] Chapter I, p. 21.
[16] P. 45.
[18] See, for instance, H. J. H. Fenton, Outlines of Chemistry, 128, 1918.
[15] Proc. Phys. Soc. Lond., 36, 94, 1924.
[17] Ap. J., 59, 197, 1924.

ionization considered by chemical theory is the separation of a *molecule* in solution into charged radicals. The essential point is the acquisition of a charge at dissociation, and this is the only feature that the chemical ionization has in common with the thermal ionization, where the *atom* is separated into a positively charged ion and an electron which constitutes the negative charge.

The step from the theory first formulated for solutions to the theory of gaseous ionization is a long one, and its legitimacy has been questioned.[19] It appears, however, that the step is justified.[20] The stellar conditions are certainly simpler than those in a solution, and if the requisite dilution obtains, the law may be expected to hold with considerable closeness. Saha contemplated pressures of the order of an atmosphere, and it may be shown that under such conditions the volume concentration would be too great and the theory would be invalid. At pressures of 10^{-4} atmospheres, however, the effect of concentration is just becoming inappreciable, and the theory probably holds with fair exactness.

LABORATORY EVIDENCE BEARING ON THE THEORY

(a) *Ultimate Lines.*[21] — The physical tests of the Saha theory that have been made in the laboratory have all supported it strongly. The fact that the ultimate lines of an atom are the lines normally absorbed by the cold vapor has long been familiar. Indeed it is this fact that is tacitly assumed in the identification of lines of zero excitation potential in the laboratory with lines which are strongest in the low-temperature furnace spectrum. De Gramont[22] designated the ultimate lines "raies de grande sensibilité" for the detection of small quantities of a substance, because they are the last to disappear from the flame spectrum when the quantity of the substance is decreased.

[19] Lindemann, quoted by Milne, Observatory, **44,** 264, 1921.
[20] Milne, Observatory, **44,** 264, 1921.
[21] Chapter VI, p. 94.
[22] C. R., **171,** 1106, 1920.

(b) *Temperature Class.* — The effect, upon the absorption spectrum of a substance, of raising the temperature has also long been recognized as an increase in the strength of lines associated with the higher excitation potentials. The use of A. S. King's "temperature class" in assigning series relations [23] involves a tacit admission of the validity of the theory of thermal ionization in predicting the relative numbers of atoms able to absorb light corresponding to different levels of energy.[24]

(c) *Furnace Experiments.* — King's explicit investigation [25] of the effects of thermal ionization in the furnace has contributed valuable positive evidence for the theory. For example, the production of the subordinate series of the neutral atoms of the alkali metals by raising the temperature was an experimental proof of the principle mentioned in the last paragraph; and the suppression of the enhanced lines of calcium by the presence of an excess of free electrons, derived from the concurrent ionization of potassium, with an ionization potential 1.77 volts lower than that of calcium, and the similar results obtained for strontium and barium, fulfill the predictions of ionization theory in a striking fashion.

(d) *Conductivities of Flames.* — The conductivity of a flame may be used as a measure of the ionization that is taking place at the temperature in question, and the available data on flame conductivities have been discussed by Noyes and Wilson [26] from the standpoint of the theory of thermal ionization. The calculations based upon the conductivities imparted to a flame by the different alkali metals, and leading to an estimate of the ionization constant, were in satisfactory agreement with the theoretical predictions of the ionization constant from the known critical potentials. The theory of thermal ionization is, therefore, strongly supported by all the laboratory investigations which have so far been undertaken in testing it.

---

[23] Russell, Ap. J., in press.
[24] A. S. King, Mt. W. Contr. 247, 1922.
[25] A. S. King, Mt. W. Contr. 233, 1922.
[26] Ap. J., **57**, 20, 1923.

SOLAR INTENSITIES AS A TEST OF IONIZATION THEORY

Before proceeding to discuss the stellar intensity curves, it is proposed to review some of the solar evidence, which can be treated as an observational test of the predictions of the theory relating to the distribution of atoms among the possible atomic states at a given temperature.

In two papers, Russell [27] has given a discussion of the solar and sunspot spectra, showing that ionization theory offers a very satisfactory interpretation of most of the observed phenomena. Attention was called to the anomalous behavior of barium and lithium,[28] and it was suggested that the theory of thermal ionization, while taking account of the temperature of the reversing layer, omitted to consider the effect of the absorption of photospheric radiation. This omission might cause a deviation such as is observed for barium, but appears inadequate to account for the behavior of lithium. In the case of lithium, low atomic weight, and a consequent high velocity of thermal agitation, has been suggested as the cause of the anomaly. The question of the absorption of photospheric radiation has more recently been discussed by Saha,[29] in the form of a correction to his own ionization equations. It has been pointed out by Woltjer [30] that the correction introduced by Saha and Swe may also be derived from considerations advanced by Einstein [31] and Milne.[32] The correction can be evaluated, but appears in every case to be rather small. The effect of the photospheric radiation is certainly one that must be included in a satisfactory theory, but at present, observation is probably not of sufficient accuracy to demand such a refinement.

The work just quoted was qualitative. A more quantitative test of ionization theory in the solar spectrum can also be made [33] by comparing the intensities of solar lines corresponding to different excitation potentials, but belonging to the same atom. The

[27] Mt. W. Contr. 225, 1922.　[28] Mt. W. Contr. 236, 1922.
[29] Saha and Swe, Nature, 115, 377, 1925.
[30] Nature, 115, 534, 1925.　[31] Phys. Zeit., 18, 121, 1917.
[32] Phil. Mag., 47, 209, 1924.　[33] Payne, Proc. N. Ac. Sci., 11, 197, 1925.

atoms which give a large number of lines in the solar spectrum are those of the first long period of the periodic table, and these, as is well known, consist of multiplets, with components of very different intensities. It appears to be legitimate to select

Atom	Excitation Potential	$\log n_r$	Intensity	Atom	Excitation Potential	$\log n_r$	Intensity
Calcium	0.00	$\bar{2}.67$	20	Chromium	0.00	$\bar{1}.20$	10
	1.88	$\bar{4}.97$	15		0.94	$\bar{2}.38$	5
	2.53	$\bar{4}.40$	8		1.02	$\bar{2}.31$	5
	2.70	$\bar{4}.26$	5		2.89	$\bar{4}.66$	2
	2.92	$\bar{4}.07$	4	Titanium	0.00	$\bar{1}.89$	5
Iron	0.00	$\bar{2}.66$	40		0.82	$\bar{2}.14$	4
	0.94	$\bar{3}.86$	30		0.90	$\bar{2}.08$	3
	1.54	$\bar{3}.31$	30		1.05	$\bar{3}.95$	3
	2.19	$\bar{4}.71$	8		1.44	$\bar{3}.81$	3
	2.46	$\bar{4}.45$	10		1.50	$\bar{3}.54$	2
	2.84	$\bar{4}.08$	8		1.87	$\bar{3}.21$	1
	2.96	$\bar{4}.02$	7		1.98	$\bar{3}.21$	1
	3.25	$\bar{5}.77$	7		2.08	$\bar{3}.04$	0
	3.38	$\bar{5}.66$	6		2.16	$\bar{4}.95$	1
	3.64	$\bar{5}.39$	8		2.24	$\bar{4}.89$	2
	4.13	$\bar{6}.93$	–		2.26	$\bar{4}.85$	0
	4.23	$\bar{6}.86$	–		2.28	$\bar{4}.83$	0
	4.35	$\bar{6}.84$	–		2.33	$\bar{4}.80$	0
	4.40	$\bar{6}.70$	–		2.39	$\bar{4}.75$	00
					2.47	$\bar{4}.67$	–
					2.56	$\bar{4}.60$	000
					2.67	$\bar{4}.52$	000

the strongest line associated with any energy level for the comparison; the strength of this line probably represents fairly well the tendency of the atom to be in the corresponding state. The atoms for which there are enough known lines of different excitation energies in the solar spectrum are those of calcium, chromium, titanium, and iron. The correlation between the excitation potential associated with a given line and the intensity of the line in the solar spectrum is illustrated by the preceding tabulation. Successive columns give the atom, the excitation potential, the computed fractional concentration, expressed logarithmically, and the observed intensity, taken from Rowland's table.

It will be seen that the correlation is very marked, and that it appears to furnish good evidence that the theory of thermal ionization predicts correctly the relative tendencies of the atoms to absorb the different frequencies. The fractional concentrations are of course not absolute values, as the number of atoms in a state of high excitation is a definite fraction, not of the *whole* number of atoms, but of the *number left over* from the lower excitations. Neither are the intensities given by Rowland absolute, and therefore the comparison appears sufficient to show the strong correlation between excitation potential and solar intensity.

CHAPTER VIII

OBSERVATIONAL MATERIAL FOR THE TEST OF IONIZATION THEORY

THE observational test of ionization theory involves a considerable program of measurement, if the accuracy necessary for a quantitative test is to be attained. The present chapter contains a synopsis of new data obtained by the writer to supplement the material already published in Harvard Circulars.[1,2] The data here presented practically complete the available material for the strong lines of known series relations in the region of the spectrum usually examined.

LINE INTENSITY

The theory predicts the degree of absorption that will be produced by each atom at a given temperature, and the related quantity that is measured is the intensity of the corresponding Fraunhofer line in the spectrum of the star. Spectrum lines are differentiated by various qualities, such as width, darkness, and wings, and their conspicuousness is governed by the intensity of the neighboring continuous background. It is not easy to specify all these quantities on an intensity scale that is one-dimensional, and the various ways in which line intensities have been estimated represent different attempts to choose and express a suitable scale.

Many of the applications of so-called line-intensity, such as the estimation of spectroscopic parallaxes, have involved ratios between the strengths of various lines in the same spectrum. This method of comparison avoids most of the difficulties caused by differences of line character and continuous background, for the lines that are to be compared are chosen because

[1] Payne, H. C. 256, 263, 1924. [2] Menzel, H. C. 258, 1924.

ESTIMATION OF INTENSITY 117

of their proximity and comparability. Harper and Young[3] have standardized the method by comparing spectrum line ratios with line ratios on an artificial scale.

Method of Estimating Intensity

In a comparison of ionization theory with observation, some measure of line-intensity is required which can be compared from class to class. It seems probable that direct estimates of intensity, for spectra of the same dispersion, density, and definition, will be comparable within the limits of accuracy of the material.

Two series of spectra were measured by the writer in order to obtain material for the test of the theory of ionization. For the first group standard lines in the spectrum of α Cygni were used for the formation of a direct intensity scale, and for the second group, comprising the cooler stars, a strip of the solar spectrum was similarly employed. An arbitrary scale was constructed by assigning a series of intensities to well placed lines in the spectrum, and using these as standards. A list of the lines used for the second group, the assigned intensity, and the intensity as given in Rowland's table, are contained in the following table.

Line	Intensity Assigned	Rowland	Line	Intensity Assigned	Rowland
4034	6	7	4046	10	30
4035	5	6	3968	13	700
4038	4	4	3934	15	1000
4064	8	20			

The estimates thus made might be defined as estimates of width-intensity-contrast between the line and the continuous background. On an ideal plate which was not burned out, such estimates would give a measure of the total energy of the line relative to the neighboring continuous spectrum. The accuracy attained by direct estimates of this kind appears to be as great as the material warrants.

[3] Harper and Young, Pub. Dom. Ap. Obs., 3, 3, 1925.

ACCURACY OF THE ESTIMATES

It is not possible at present to evaluate the accuracy of these estimates with the same precision as for other physical quantities, but the consistency of the readings from comparable plates of the same star will at least give a measure of the value of the estimates. Table XVIII contains the measures on forty-three lines in the spectrum of β Gruis, taken from six plates of the same dispersion, and comparable quality, density, and definition. Successive columns give the wave-length, the arithmetic mean

TABLE XVIII

Line	Int.	σ	Line	Int.	σ	Line	Int.	σ	Line	Int.	σ
4215	8.4	1.0	4319	4.8	0.8	4376	7.8	0.6	4451	4.5	1.5
4227	16.0	1.4	4321	4.0	0.0	4379	4.2	0.3	4455	5.0	1.0
4247	6.0	0.0	4326	10.4	0.5	4383	10.3	0.8	4462	6.0	0.8
4250	7.0	0.0	4330	4.6	0.8	4395	6.3	0.4	4482	7.7	0.7
4254	9.0	1.0	4332	3.6	0.5	4398	2.7	0.8	4490	7.3	0.4
4260	9.0	2.1	4333	4.0	0.0	4402	6.3	0.4	4495	7.5	0.4
4272	8.7	0.8	4337	8.7	0.8	4405	9.0	0.5	4502	6.0	0.0
4275	9.5	1.3	4340	9.5	1.1	4409	9.0	0.5	4554	5.3	0.4
4283	4.3	0.4	4352	9.2	1.1	4415	7.7	0.8	4564	5.8	0.7
4290	10.6	1.0	4360	6.8	1.0	4435	9.2	0.9	4572	6.0	1.0
4315	8.3	0.7	4370	6.8	0.6	4444	8.9	1.2			

intensity, and the standard deviation σ. These measures are strictly representative of the material as a whole, for the plates of β Gruis were measured at wide intervals in the ordinary course of the work, and were selected for illustration because there was a greater number of suitable plates of this star than for any other.

HOMOGENEITY OF MATERIAL

The observational material on line-intensities follows in tabular form. The measures were made in two groups, comprising respectively the hotter stars and the stars cooler than Class A0, and different intensity scales were used for the two. The solar scale mentioned above was used for the second group of stars;

the first group was referred to standard lines in the spectrum of α Cygni. The distribution of the stars in the two groups among the spectral classes was as follows:

	Group I	Bo	B1	B2	B3	B5	B8	B9	Ao
giants		4	7	7	6	8	6	3	17
super-giants		-	-	-	-	2	1	-	-

	Group II	B9	Ao	A2	A3	A5	Fo	F2	F5	F8	Go	G5	Ko	K2	K5	Ma	Mb	Md
dwarfs		-	-	-	-	-	4	1	2	3	2	2	-	-	-	-	-	-
giants		1	9	3	5	2	-	-	-	-	3	5	20	4	8	1	5	1
super-giants		-	-	-	-	-	3	1	5	1	2	-	2?	1?	1?	3	1	-

If it were possible to use a series of giants throughout, the task of determining the intensity maxima would be greatly simplified. Among the hotter stars the differences introduced by absolute magnitude are not great enough to make the maxima difficult to determine. With later classes, however, the changes with absolute magnitude are very marked. As will be pointed out in an ensuing chapter,[4] the actual strength of the lines differs considerably from giant to dwarf, owing to the difference in the effective optical depth of the photosphere. This difference in strength is in addition to the well-known " absolute magnitude effect " which is shown, for example, by the enhanced lines; it increases the difficulty of making estimates of line change from one class to the next, since, owing to selection, the available stars are far from homogeneous in absolute magnitude. In addition to this factor, there is the practical difficulty of making comparable estimates on the sharp narrow lines of a supergiant and those of a dwarf, since the lines of a dwarf tend to be hazy and lack contrast with the background.

It might be expected, from the distribution in luminosity of the stars used, that irregularities in the intensity sequence would probably occur in the F classes and at Ma. For the purpose of estimation of maxima, the F classes are not of very great importance, as few of the maxima under present investigation occur there, but the irregularity at Ma may well prove to be serious.

[4] Chapter X, p. 142.

There is indeed a general tendency for the intensity of metallic lines to increase at Ma. All the Ma stars measured were of very high luminosity, and probably the rise of intensity is due to this feature, or rather to the increase of material above the photosphere that accompanies it. A maximum is only assumed to occur at Ma when a line increases regularly through the K types, as do the lines of neutral calcium. The iron and titanium maxima obviously occur earlier in the sequence, although the lines of both these elements are often noticeably strengthened at Ma.

The following tabulation contains the data on line-intensity for all the lines of known series relations that have been measured up to the present. All the measures were made by the writer, excepting those for zinc, which are taken from Menzel's paper.[5] Successive columns of the table contain the atom, the series relations, the wave-length, and the observed intensities in the various spectral classes. The column headed "Blends" is a direct transcription from Rowland's tables, and contains details both of the line under consideration and of closely adjacent lines. The column headed "Remarks" contains the writer's own conclusions, based on solar evidence, astrophysical behavior, and laboratory affinities, as to the source and maximum of the line that has been measured.

The recorded intensities, for classes cooler than Bo, are derived from the selection of stars mentioned earlier in the present chapter. A list of the individual stars is contained in Appendix III. Four typical O stars have been selected to represent that class. The figures in the final column refer to the notes to the table, which are listed under the respective atoms, and give the observed maximum, the intensities and origins of blended lines (in Rowland's notation), and short remarks, which indicate whether or no the observed behavior is to be attributed to the line considered. Maxima that are obviously due to another line are placed in parentheses.

[5] Menzel, H. C. 258, 1924.

TABLE XIX

Atom	λ	Series	B9	Ao	A2	A3	A5	Fo	F5	F8	Go	G5	Ko	K2	K5	Ma	Mb	Notes
H	3970.1	2P – 7D	20.0	17.6	20.0	15.6	15.0	17.2	17.8	..	18.0	20.0	24.5	..	..	30.0	..	1
	4101.7	2P – 6D	18.0	16.0	16.3	13.6	15.0	13.9	10.7	10.6	9.4	7.0	7.0	7.0	7.3	9.0	6.0	2
	4340.5	2P – 5D	..	16.0	14.3	12.2	14.6	13.2	10.8	9.2	9.4	9.0	8.7	8.4	9.2	9.6	9.0	3
	4861.3	2P – 4D	15.0	14.0	14.0	14.6	13.3	12.3	11.0	9.0	8.5	7.6	6.6	5.6	5.1	6.7	4.0	4
			ζ Pup	δ Cir	29 CMa	τ CMa	Bo	B1	B2	B3	B5	B8	B9	Ao	A2	A3	A5	
He	4713.4	1²P – 3³S	..	..	..	..	6.5	8.2	7.5	6.0	6.7	4.2	0.0	0.0	0.0	0.0	0.0	1
	4713.1	1²P – 3³S																
	4121.0	1²P – 4²S	..	..	..	..	8.5	9.5	11.0	9.2	6.4	4.2	0.0	0.0	0.0	0.0	0.0	2
	4120.9	1²P – 4²S																
	4471.7	1²P – 3²D	..	6.5	8.5	8.0	11.0	11.5	11.6	11.8	11.1	9.7	8.0	0.0	0.0	0.0	0.0	3
	4471.5	1²P – 3²D																
	4026.4	1²P – 4²D	4.0	6.9	9.0	8.0	12.0	12.7	14.0	15.4	12.0	10.8	8.5	0.0	0.0	0.0	0.0	4
	4026.2	1²P – 4²D																
	4921.9	1P – 3D	..	..	..	..	10.0	12.4	10.7	10.0	10.0	7.0	4.0	0.0	0.0	0.0	0.0	5
	4387.9	1P – 4D	..	..	4.0	4.0	10.0	10.3	11.0	11.5	9.2	..	4.0	0.0	0.0	0.0	0.0	6
	4143.8	1P – 5D	..	..	5.0	4.0	9.6	10.0	10.7	12.0	7.5	4.9	3.5	0.0	0.0	0.0	0.0	7
	4009.3	1P – 6D	..	..	..	..	7.0	9.1	10.2	11.4	5.8	4.0	0.0	0.0	0.0	0.0	0.0	8
He+	4685.8	3D – 4F	em.	5.8	em.	6.0	4.0	0.0	0.0	0.0	0.0	0.0	0.0	0.0	0.0	0.0	0.0	9
	4541.6	4F – 9G	6.0	5.3	5.5	6.0	4.0	0.0	0.0	0.0	0.0	0.0	0.0	0.0	0.0	0.0	0.0	10
	4199.9	4F – 11G	5.0	3.5	6.1	5.0	0.0	0.0	0.0	0.0	0.0	0.0	0.0	0.0	0.0	0.0	0.0	11
	4025.6	4F – 13G	4.0	6.9	9.0	8.0	0.0	0.0	0.0	0.0	0.0	0.0	0.0	0.0	0.0	0.0	0.0	12
C+	4267	3²D – ²F	0.0	0.0	0.0	0.0	5.0	7.4	7.7	8.0	7.8	4.5	3.0	0.0	0.0	0.0	0.0	1

TABLE XIX (continued)

Atom	λ	Series	B9	A0	A2	A3	A5	F0	F5	F8	G0	G5	K0	K2	K5	Ma	Mb	Notes
Mg	5183.7	$1^3P - 1^3S$	..	..	..	..	..	8.0	..	..	8.0	8.0	8.0	10.0	8.0	10.0	..	1
	5172.7																	
	5167.4																	
	4571.1	$1S - 1^3P$	..	..	..	..	2.0	3.5	5.2	3.3	6.4	5.8	6.2	6.8	6.9	7.0	6.1	2
	4351.9	$1P - 5D$	..	2.0	4.0	5.0	4.3	6.3	7.4	5.0	7.3	8.1	8.0	7.0	8.1	8.6	9.0	3
Mg+	4481.3	$2^2D - 3^2F$	5.0	4.6	6.0	5.5	6.7	8.0	7.2	8.1	8.3	8.6	9.0	7.7	8.0	9.4	7.6	4
	4481.1	$2^2D - 3^2F$																
Al	3961.5	$1^2P - 1^2S$	..	tr.	2.0	5.3	..	5.7	5.5	..	8.3	8.0	8.5	..	..	9.0	11.0	1
	3944.0	$1^2P - 1^2S$	..	tr.	2.0	6.0	..	5.2	6.0	8.0	8.0	8.3	8.5	..	.	8.5	11.0	2
Si	3905		..	..	2.0	..	4.0	8.8	9.3	11.5	11.7	11.4	11.3	10.0	10.0	9.6	8.6	1
			ζ Pup	δ Cir	29 CMa	τ CMa	B0	B1	B2	B3	B5	B8	B9	A0	A2	A3	A5	
Si+	4131		..	..	..	..	2.0	3.0	2.7	3.5	4.4	3.6	6.2	9.3	7.0	..	..	2
	4128		..	..	..	..	2.0	3.0	2.7	3.5	4.4	3.6	6.2	9.3	7.0	..	..	
Si++	4574		0.0	0.0	0.0	0.0	3.0	8.0	8.0	2.0	0.0	0.0	0.0	0.0	0.0	0.0	0.0	3
	4568		0.0	0.0	0.0	0.0	2.0	9.0	9.0	4.0	0.0	0.0	0.0	0.0	0.0	0.0	0.0	4
	4552		0.0	0.0	0.0	0.0	5.5	10.0	10.0	5.0	0.0	0.0	0.0	0.0	0.0	0.0	0.0	5
Si+++	4116		..	..	5.0	6.0	8.3	4.7	0.0	0.0	0.0	0.0	0.0	0.0	0.0	0.0	0.0	6
	4096		..	..	8.7	6.0	9.7	5.2	3.6	0.0	0.0	0.0	0.0	0.0	0.0	0.0	0.0	7
	4089		..	..	7.5	8.0	9.2	5.5	5.2	0.0	0.0	0.0	0.0	0.0	0.0	0.0	0.0	8
			B9	A0	A2	A3	A5	F0	F5	F8	G0	G5	K0	K2	K5	Ma	Mb	
Ca	4581.4	$1^3D - 3^3F$	0.0	0.0	0.0	3.0	2.5	4.2	6.7	5.0	7.7	8.1	7.8	8.0	7.1	8.3	6.2	1
	4454.8	$1^3P - 2^3D$	0.0	0.0	0.0	3.0	5.0	2.6	5.0	6.2	6.0	5.1	5.0	4.8	4.4	5.6	5.0	2
	4434.9	$1^3P - 2^3D$	0.0	0.0	0.0	4.0	4.3	5.2	6.2	6.0	7.5	7.6	7.9	9.2	8.8	9.7	9.3	3

TABLE XIX (continued)

Atom	λ	Series	B9	A0	A2	A3	A5	F0	F5	F8	G0	G5	K0	K2	K5	Ma	Mb	Notes
Ca	4307.7	$1^3P - 3^3P'$	0.0	3.3	3.0	4.4	3.6	4.9	6.5	7.2	9.5	8.6	8.6	10.3	10.5	12.0	13.1	4
	4302.5	$1^3P - 3^3P'$	0.0	0.0	0.0	. .	8.0	. .	3.5	. .	. .	5.5	. .	4.0	4.5	5.0	8.0	5
	4299.0	$1^3P - 3^3P'$	0.0	0.0	3.0	5.2	5.0	6.8	7.4	6.5	6.6	8.0	8.0	8.6	7.9	. .	6.0	6
	4289.4	$1^3P - 3^3P'$	0.0	0.0	3.0	4.4	4.6	7.6	7.8	6.7	8.2	7.4	7.7	8.6	9.3	11.4	10.5	7
	4283.0	$1^3P - 3^3P'$	0.0	0.0	0.0	. .	5.0	3.0	. .	. .	4.0	5.0	6.1	5.0	6.0	8.0	4.4	8
	4226.7	$1S - 1P$	3.0	2.3	3.0	5.8	6.3	7.9	9.3	8.6	10.4	9.7	11.7	13.6	14.5	14.2	16.0	9
Ca+	3968.5	$1^2S - 1^2P$	20.0	17.6	20.0	15.6	15.0	17.2	17.8	20?	18.0	20.0	24.5	. .	. .	30.0	. .	10
	3933.7	$1^2S - 1^2P$	5.0	10.3	13.3	13.8	15.0	17.1	20.0	20.0	19.6	20.0	21.5	. .	. .	30.0	. .	11
Sc+	4246.8	$3F - 3D'$	0.0	3.0	2.0	6.0	5.0	6.0	7.0	5.0	6.3	5.6	5.3	5.6	5.4	8.6	6.8	1
	4320.8	$3F - 3D'$	0.0	0.0	0.0	0.0	0.0	3.8	5.5	4.0	4.0	3.5	3.8	4.5	4.0	5.0	4.0	2
Ti	4395.2	$1^5D - 5F$	0.0	0.0	0.0	3.0	. .	6.7	7.2	. .	5.7	5.7	5.4	6.2	6.8	7.2	6.5	1
	4301.1	$1^5F - 5D$	0.0	0.0	0.0	0.0	8.0	. .	3.5	. .	. .	5.5	. .	4.0	4.5	5.0	8.0	2
	4300.9	$1^5F - 5D$	0.0	3.0	3.0	. .	. .	6.0	7.6	7.0	5.0	6.5	5.0	8.0	5.6	10.3	13.0	3
	4298.7	$1^5F - 5D$	0.0	0.0	3.0	5.2	5.0	6.8	7.4	6.5	6.6	8.0	8.0	8.6	7.9	. .	6.0	4
	4290.9	$1^5F - 5D$	0.0	0.0	3.0	4.4	4.6	7.6	7.8	6.7	8.2	7.4	7.7	8.6	9.3	11.4	10.5	5
	4289.1	$1^5F - 5D$	0.0	0.0	1.0	3.0	. .	3.5	5.2	5.0	6.2	7.2	7.6	9.0	8.4	9.5	9.4	6
	4274.6	$1^5F - 5D$	0.0	0.0	1.5	4.5	6.0	5.6	7.0	6.0	7.0	7.0	6.5	. .	7.0	8.0	. .	7
	3998.7	$1^3F - 3F_x$	0.0	0.0	0.0	0.0	2.0	3.5	5.2	3.3	6.4	5.8	6.2	6.8	6.9	7.0	6.1	8
Ti+	4571.9	$1^2H - 1^2G'$	0.0	0.0	0.0	0.0	2.0	2.5	5.2	3.3	6.6	6.4	6.7	7.2	6.9	7.0	5.6	9
	4563.8	$1^2P - 1^2D$	0.0	0.0	0.0	0.0	0.0	4.0	6.6	5.3	6.8	6.0	6.4	6.8	7.0	6.6	6.0	10
	4501.3	$1^2G - 1^2F'$	0.0	0.0	0.0	5.0	8.0	6.9	8.1	7.8	7.5	7.6	7.9	9.2	8.6	8.0	9.0	11
	4443.8	$2D - 2F$	0.0	0.0	0.0	3.0	. .	6.7	7.2	. .	5.7	5.7	5.4	6.2	6.8	6.0	6.5	12
	4395.0	$2D - 2F$																

123

TABLE XIX (continued)

Atom	λ	Series	B9	A0	A2	A3	A5	F0	F5	F8	G0	G5	K0	K2	K5	Ma	Mb	Notes
Ti+	4337.9	$^2D - ^2P_x$	0.0	0.0	0.0	0.0	0.0	5.0	6.3	5.0	7.2	7.8	7.2	7.2	8.9	9.0	8.3	13
	4315.0	$^4P - ^4D$	0.0	3.0	3.5	5.0	4.6	6.9	7.5	4.6	6.6	5.6	6.1	7.0	7.0	8.0	8.1	14
	4307.9	$^4P - ^4D$	0.0	3.3	3.0	4.4	3.6	4.9	6.5	7.2	9.5	8.6	8.6	10.3	10.5	12.0	13.1	15
	4301.9	$^4P - ^4D$	0.0	0.0	0.0	0.0	8.0	::	3.5	::	::	5.5	::	4.0	4.5	5.0	8.0	16
	4300.0	$^4P - ^4D$	0.0	3.0	3.0	::	::	6.0	7.6	7.0	5.0	6.5	5.0	8.0	5.6	10.0	13.0	17
	4290.2	$^4P - ^4D$	0.0	0.0	3.0	4.4	4.6	7.6	7.8	6.7	8.2	7.4	7.7	8.6	9.3	11.0	10.5	18
V	4395.2	$^6D - ^6F$	0.0	0.0	0.0	3.0	0.0	6.7	7.2	::	5.7	5.7	5.4	6.2	6.8	7.2	6.5	1
	4379.2	$^6D - ^6F$	0.0	0.0	0.0	0.0	0.0	3.0	2.5	3.0	3.7	2.5	2.8	3.5	3.0	4.0	4.2	2
	4332.8	$^4G - ^4G_x$	0.0	0.0	0.0	0.0	0.0	4.0	::	::	4.5	5.0	4.5	5.0	4.0	5.0	4.2	3
	4330.1	$^4G - ^4G_x$	0.0	0.0	0.0	0.0	0.0	0.0	0.0	0.0	4.5	5.0	4.5	4.5	4.0	3.5	3.8	4
Cr	4359.8	$^5D - ^5F$	0.0	2.0	::	3.0	::	3.6	5.6	5.0	6.8	6.2	6.3	6.8	7.1	6.0	6.6	1
	4351.9	$^5D - ^5F$	0.0	2.0	4.0	5.0	4.3	6.3	7.4	5.0	7.3	8.1	8.0	7.0	8.1	8.0	9.0	2
	4289.7	$^5S - ^5P$	0.0	0.0	3.0	4.4	4.6	7.6	7.8	6.7	8.2	7.4	7.7	8.6	9.3	11.4	10.5	3
	4274.9	$^5S - ^5P$	0.0	0.0	1.0	3.0	::	3.5	5.2	5.0	6.2	7.2	7.6	9.0	8.4	9.5	9.4	4
	4254.4	$^5S - ^5P$	0.0	0.0	0.0	2.5	4.0	3.3	4.6	5.0	6.4	8.0	8.0	8.6	8.6	9.5	9.9	5
Mn	4451.6	$^1D - ^4D$	0.0	2.0	3.0	::	3.0	2.0	::	::	6.0	4.0	3.5	5.0	::	5.0	5.0	1
	4414.9	$^1D - ^4D$	3.0	2.3	3.3	4.8	4.6	6.4	8.0	7.2	7.0	6.5	7.2	7.6	7.4	7.0	8.0	2
	4041.4	$^1D - ^6D_x$	0.0	0.0	0.0	2.5	2.0	1.9	5.2	3.5	6.0	4.8	4.8	5.5	6.0	6.0	6.0	3
	4036.5	$^1S - ^1P$	0.0	::	::	::	::	::	::	::	::	3.3	4.0	::	::	::	::	4
	4034.5	$^1S - ^1P$	0.0	::	::	::	::	::	::	::	5.0	3.6	4.0	::	6.0	::	::	5
	4033.1	$^1S - ^1P$	0.0	2.0	1.0	3.6	4.0	4.0	5.0	6.0	6.0	4.3	5.0	7.0	7.0	8.5	8.0	6
	4030.8	$^1S - ^1P$	0.0	::	::	4.3	::	4.3	5.7	5.5	7.0	5.4	6.4	8.0	8.0	6.0	5.5	7
	4068	unclas.		::	::		::	::	7.6	5.0	6.6	5.7	5.8	8.0	8.5	6.0	5.5	8
	4092	unclas.		::	::	3.0	::	::	::	3.0	4.0	4.2	4.2	5.0	5.0	6.0	5.5	9

TABLE XIX (continued)

Atom	λ	Series	B9	Ao	A2	A3	A5	Fo	F5	F8	Go	G5	Ko	K2	K5	Ma	Mb	Notes
Fe	4489.7	$1^5D - 1^7F$	0.0	0.0	2.0	0.0	0.0	4.2	7.2	6.3	6.8	6.8	7.6	8.4	7.4	8.0	7.6	1
	4482.3	$1^5D - 1^7F$	0.0	4.0	4.6	6.0	5.5	6.7	8.0	7.2	8.1	8.3	8.6	9.0	7.7	8.0	7.6	2
	4461.7	$1^5D - 1^7F$	0.0	0.0	0.0	0.0	3.0	2.0	4.2	5.0	6.0	6.0	6.0	6.3	6.5	7.0	7.0	3
	4375.9	$1^5D - 1^7F$	0.0	0.0	2.5	4.4	5.5	6.7	7.1	: :	7.8	6.7	6.9	: :	8.0	7.0	8.8	4
	4216.1	$1^5D - 1^7P$	0.0	2.0	1.5	3.8	3.3	5.6	6.6	2.6	8.6	7.3	8.5	9.4	8.1	8.0	7.9	5
	4415.1	$1^3F - 3D$	3.0	2.3	3.3	4.8	4.6	6.4	8.0	7.2	7.0	6.5	7.2	7.6	7.4	7.0	8.0	6
	4404.7	$1^3F - 3D$	0.0	0.0	0.0	0.0	6.0	4.4	6.7	5.0	6.1	7.4	7.8	8.8	8.0	8.0	9.1	7
	4383.4	$1^3F - 3D$	3.0	2.5	4.0	5.4	4.6	6.7	7.0	8.2	9.1	10.3	10.3	10.5	9.9	11.0	10.3	8
	4337.0	$1^3F - 3D$	0.0	0.0	0.0	0.0	0.0	5.0	6.3	5.0	7.2	7.8	7.2	7.2	8.9	9.0	8.3	9
	4291.4	$1^3F - 3D$	0.0	0.0	3.0	4.4	4.6	7.6	7.8	6.7	8.2	7.4	7.7	8.6	9.3	11.0	10.5	10
	4325.8	$1^3F - 3G$	0.0	0.0	3.5	5.5	3.3	6.3	7.8	9.0	10.0	11.0	11.3	11.7	10.9	11.0	10.2	11
	4307.9	$1^3F - 3G$	0.0	3.3	3.0	4.0	3.6	4.9	6.5	7.2	9.5	8.6	8.6	10.3	10.5	12.0	13.1	12
	4271.8	$1^3F - 3G$	0.0	2.3	2.5	4.2	4.5	4.5	6.5	6.3	7.2	8.3	8.6	9.2	8.7	10.0	9.1	13
	4260.5	$1^3F - 3G$	0.0	3.0	1.5	4.2	4.0	: :	5.5	6.6	7.0	8.0	8.8	9.0	8.1	10.0	9.0	14
	4290.8	$1^3F - 3G$	0.0	0.0	2.0	3.6	3.3	5.6	5.3	6.3	6.4	7.8	7.9	8.4	8.0	9.0	7.6	15
	4143.9	$1^3F - 3F$	0.0	0.0	2.5	4.4	4.0	4.8	6.1	5.7	7.7	7.7	8.9	8.6	8.5	11.0	10.0	16
	4132.	$1^3F - 3F$	0.0	2.0	2.0	4.0	3.3	4.2	7.0	4.0	6.0	5.2	5.5	6.0	6.5	5.0	4.0	17
	4071.7	$1^3F - 3F$	0.0	3.6	2.0	4.0	2.0	4.3	6.6	5.7	7.8	7.5	9.2	9.0	9.0	9.5	8.6	18
	4063.6	$1^3F - 3F$	0.0	2.0	2.0	3.6	3.0	4.8	5.8	5.6	7.2	7.5	8.0	9.0	8.0	9.5	9.0	19
	4045.8	$1^3F - 3F$	0.0	2.0	2.5	4.6	4.0	5.6	6.9	7.6	8.8	9.2	10.3	10.6	8.6	11.0	10.8	20
	4005.2	$1^3F - 3F$	0.0	0.0	2.5	4.6	5.0	5.0	6.3	6.0	8.3	7.2	6.6	9.0	8.0	7.0	: :	21
	4299.2	$1^7D - 7D$	0.0	0.0	3.0	: :	: :	6.0	7.6	7.0	5.0	6.5	5.0	9.0	5.6	10.3	13.0	22
	4271.2	$1^7D - 7D$	0.0	2.3	2.5	4.2	4.5	4.2	6.5	6.3	7.2	8.3	8.6	9.2	8.7	10.0	9.1	23
	4260.5	$1^7D - 7D$	0.0	3.0	1.5	4.2	4.0	: :	5.5	6.6	7.0	8.0	8.0	9.0	8.1	8.5	9.0	24
	4250.1	$1^7D - 7D$	0.0	0.0	2.0	3.6	4.0	5.6	5.3	6.3	6.4	7.8	7.9	8.4	8.0	9.0	7.6	25
	4187.8	$1^7D - 7D$	0.0	0.0	2.0	5.0	6.0	: :	: :	: :	6.0	5.5	6.0	: :	5.6	: :	8.0	26
	4482.3	$1^5P - 3^5D'$	0.0	4.0	4.6	6.0	5.5	6.7	8.0	7.2	8.1	8.3	8.6	9.0	8.7	8.0	7.6	27
	4408.4	$1^5P - 3^5D'$	0.0	0.0	0.0	6.0	4.0	3.5	5.7	4.0	5.6	6.0	6.0	8.0	7.9	7.0	9.0	28

TABLE XIX (continued)

Atom	λ	Series	B9	A0	A2	A3	A5	F0	F5	F8	G0	G5	K0	K2	K5	Ma	Mb	Notes
Fe	4352.7	1⁵P — 1⁵S′	0.0	2.0	4.0	5.0	4.3	6.3	7.4	5.0	7.3	8.1	8.0	7.0	8.1	8.0	9.0	29
	4315.1	1⁵P — 1⁵S′	0.0	3.0	3.5	5.0	4.6	6.9	7.5	4.6	6.6	5.6	6.1	7.0	7.0	8.0	8.1	30
	4282.4	1⁵P — 1⁵S′	0.0	0.0	0.0	0.0	5.0	3.0	..	..	4.0	5.0	6.1	5.0	6.0	8.0	4.4	31
	4258.4	1⁵D — 1⁵P′	0.0	0.0	0.0	0.0	0.0	4.0	4.8	..	4.5	5.2	4.3	4.5	5.3	10.0	4.0	32
	4216.2	1⁵D — 1⁵P′	0.0	2.0	1.5	3.8	3.3	5.6	6.6	2.6	8.6	7.3	8.5	9.4	8.1	9.5	7.9	33
	4134.3	1⁵D — 1⁵P′	0.0	0.0	0.0	3.0	2.0	4.0	3.5	4.5	5.5	5.2	5.6	7.0	7.0	5.0	5.5	34
	3953	unclas.	0.0	0.0	2.0	5.3	..	5.7	5.5	..	8.0	8.0	7.5	..	..	..	..	35
	3999	unclas.	0.0	0.0	1.5	4.5	6.0	5.6	7.0	6.0	7.6	7.0	6.5	..	7.0	8.0	..	36
	4172	unclas.	0.0	2.0	3.0	4.6	4.6	6.3	9.0	5.7	7.3	5.8	6.4	7.3	6.0	9.0	8.2	37
	4401	unclas.	0.0	0.0	0.0	3.0	6.0	6.6	8.0	6.2	6.8	6.0	5.5	6.2	6.9	6.0	6.5	38
	4462	unclas.	0.0	0.0	0.0	0.0	3.0	2.0	4.2	5.0	6.0	6.0	6.0	6.3	6.5	7.2	7.0	39
	4476	unclas.	0.0	0.0	0.0	3.0	5.0	5.2	..	..	6.3	4.0	4.1	4.0	4.3	7.5	4.7	40
Fe+	4173.3	2⁴P — 1⁴D′	0.0	2.0	3.0	4.6	4.6	6.3	9.0	5.7	7.3	5.8	6.4	7.3	6.0	8.0	8.2	41
	4178.8	2⁴P — 1⁴F	0.0	0.0	2.5	4.3	..	6.8	9.1	..	6.0	..	..	..	..	..	6.5	42
	4416.8	2⁴P — 1⁴D′	0.0	2.3	3.3	4.8	4.6	6.4	8.0	7.2	7.0	6.5	7.2	7.6	7.4	7.0	8.0	43
Zn	4810.5	1³P — 1³S	0.0	0.0	0.0	0.0	0.0	tr	..	tr	1	tr	0	..	0	..	..	1
	4722.2	1³P — 1³S	0.0	0.0	0.0	0.0	tr	tr	..	tr	1	tr	1—	..	1—	..	..	2
Sr	4607.3	1S — 1P	0.0	0.0	0.0	0.0	0.0	2.0	4.0	..	7.0	..	8.5	8.0	8.7	9.0	9.2	1
Sr+	4215.5	1²S — 1²P	..	2.0	1.5	3.8	3.3	5.6	6.6	2.6	8.6	5.3	8.5	9.4	8.1	9.0	7.9	2
	4077.7	1²S — 1²P	..	4.2	2.5	4.2	5.0	6.9	8.4	8.6	9.2	7.8	9.5	9.3	8.3	11.0	10.8	3
Y+	4374.9		0.0	0.0	2.5	2.4	5.5	6.7	7.1	..	7.8	6.7	6.9	..	8.0	8.8	9.6	1
	4177.5		0.0	0.0	2.5	4.3	..	6.8	9.1	..	6.0	..	..	..	..	..	6.5	2
	4398.	³D — ³P	..	..	..	..	..	..	..	..	7.0	..	..	4.0	2.6	2.0	3.7	3
Ba+	4554	1²S — 1²P	..	0.0	2.0	..	..	4.0	4.7	..	3.5	3.0	4.6	4.8	5.5	5.5	5.6	1

NOTES ON OBSERVATIONAL MATERIAL

Notes to Table XIX

Atom	Note	Max.	Blends	Remarks
H	1	Ao	..	No measures available across the whole range of these lines. They are blended with He+ in the O types. For a discussion of the maximum of these lines, see p. 166
	2	Ao	..	
	3	Ao	..	
	4	Ao	..	
He	1–8	B3	..	Maximum well determined. Unblended
	4	B3	He+	See Note 12
	9–12	O		Probably blended. See H. C. 263, 1924
C	1	B3	..	Unblended
Mg	1	K2?	..	Effectively unblended. Material very meager
	2	K5	..	Unblended
	3	none	Fe 2; Cr 5; Mg 5	Cr probably predominates
	4	A2	–, Fe 5; Fe 3	Fe predominates at lower temperature; Mg probably responsible for maximum
Al	1	none	..	
	2	none	..	
Si	1	Go	–2; –1; Si 12; –2; –1	Si predominates, and is responsible for maximum
	2	Ao	..	
	3–5	B1–B2	..	
	6	Bo–O	..	
	7	Bo–O	N++	
	8	Bo–O		
Ca	1	Ma	Ca 4; Co; Fe 4	Ca probably responsible for rise at Ma
	2	Ma	Ca, Zr 5; Mn 1; Mn, Ti 2; Mn 2; Ca 2	Ca probably predominates. Enhanced line suspected near Go
	3	Ma	Ca 5; Fe 2; Ca 4	Calcium predominates. Enhanced line suspected near Go
	4	none	Ca 3; Fe 6	In G band
	5	none	Ti 2; Fe 2; –2; Ca 4; –2	Maximum undetermined. In G band
	6	K2	Ti 2; –2; Ca 3; –1; Ti, Fe 4	Fe probably responsible for maximum
	7	Ma?	Ti 2; Ca 4; Ti 2; Cr 5; Ti; Fe 1	Chromium (ultimate) line probably obliterates the Ca line. Maximum at Ma due to Ca?
	8	Ma	Fe 5; Ca 4	
	9	none		Unblended
	10	?		Hydrogen predominates before A3

128   OBSERVATIONAL MATERIAL

Atom	Note	Max.	Blends	Remarks
	11	?		Unblended
Sc	1	F5	Y? 5; Fe 4	Sc predominates, at least at maximum
	2	F5	Sc 3; −2	Sc predominates
Ti	1	K5	Ti 3; V, Zr 2	Blended with Ti+. See Note 10
	2	(none)	Ti 2; Fe 2; Ca 4; −2	Ca causes rise at Ma. Ti obliterated
	3	K2	Ti 3	Ti+ causes rise at F5. Rise at Ma unexplained
	4	K2	Ti 2; Ca 3; Ti, Fe 4	
	5	(Ma)	Ti 2; Ca 4; Cr 5; Ti 1; Ti 2; Fe 1	Ca and Cr cause rise at Ma. Ti obliterated
	6	(none)	Ti 2; Cr 7	Cr (ultimate) line predominates
	7	none	Fe 4; Co 4; Fe 4; Ti 4	Possibly an enhanced line accounts for maximum near F5?
	8	Go (F5?)	Mg 5; Ti 6	Mg accounts for maximum at Ma
	9	Go (F5?)	Ti 4	Unblended
	10	Go (F5?)	Ti, −5	Probably unblended
	11	F5	Fe 3; Ti 5	Maximum at K2 due to Fe
	12	F5?	Ti 3; V, Zr 2	Blended with Ti. See Note 1
	13	?	Fe 5; Cr 3; Ti 4	Fe predominates. Maximum uncertain
	14	F5	Ti 3; Fe 4	Rise at Ma due to Fe. In G band
	15	Go	Ca 3; Fe 6	Rowland gives no Ti. Other lines account for later maximum. In G band
	16	?	Ti 2; Fe 2; Ca 2	Maximum undetermined. In G band
	17	F5	Ti 3	Unblended
	18	Go (F5?)	Cr 5; Ti 1; Ti 2	Cr accounts for strength in Ma
V	1	K5	Ti 3; V, Zr 2	Ti and Ti+ lines blended. V probably obliterated
	2	none	V 4	Unblended
	3	none	Ti, Ni 2; V o	V probably effective at low temperatures, as these are the ultimate lines
	4	none	La 1 N; V o	
Cr	1	K5	Cr 3	Unblended
	2	none	Mg 5 Nd? Cr 5; Fe 4	
	3	none	Ca 4; Cr 5; Ti 1; Ti 3	Cr probably predominates
	4	none	Ti 2; Cr 7d?	Cr predominates
	5	none	Cr 8	Unblended
Mn	1	none	Mn 3	Unblended
	2	none	Mn 2; Fe 8	Fe predominates?
	3	none	Fe 3; Mn 5; Zr, −1	
	4	none	Co 2; Mn 4d?	Mn predominates
	5	none	Mn−Fe 6d?	
	6	none	Fe−Mn 7d?	

NOTES ON OBSERVATIONAL MATERIAL 129

Atom	Note	Max.	Blends	Remarks
	7	Ma	Fe, Ti 5; Mn 4; Mn 5	Mn predominates?
	8	K5	Fe—Mn 6	
	9	Ma	Fe 2; Co, Mn 3; Fe 1; V, Ca 3d?	
Fe	1	K2	Mn—Fe	Effectively unblended
	2	(K2)	Fe 3, Fe 5	No. 2 the weaker line, Mn+ affects the line at and before A5, producing maximum at A3. See No. 27
	3	none	Fe—Mn, 3 Nd?	Effectively unblended
	4	none	Sc, Fe? 3; Zr o; V, Mn 2	Y+ accounts for maximum at Go
	5	-	Sr+	Due entirely to Sr+
	6	K2?	Mn 2; Fe 8; −3	Maximum at F5 due to Fe+
	7	K2	Fe 10	Unblended
	8	K2	Fe 15	Unblended
	9	(Ma)	Fe 5; Cr 3; Ti 4	Ti+ ($^2D - {}^2P$) causes maximum at F5; possibly Cr ($^5D - {}^5F$) causes rise at Ma
	10	(Ma)	Ti 3; Ti 2; Fe 2	Ti predominates?
	11	K2	Fe 8; Sc 4; Ti, Zr 1 Ni, Cr 1	Fe probably predominates
	12	(none)	Ca 3; Fe 6	Ca produces rise in late classes?
	13	K2	Fe 15	Unblended. Rise at Ma unexplained
	14	K2	Fe 3 d? Fe 10	Unblended. Rise at Ma unexplained, unless due to second Fe line
	15	K2	Fe 8; Fe 8	Rise at Ma unexplained. See No. 25
	16	Ko	Fe 15; Fe 4	
	17	K5	Fe 10; −3	Maximum at K5 due to unknown line?
	18	Ko	Fe 1; Fe 15	
	19	K2	Fe 4; Fe 20	
	20	K2	Mn 2, 1; Co 5; Fe 30	Fe predominates
	21	K2	Fe 7; ? 3; Fe 1; −1, 1, 1	
	22	K2	Ca 3; Ti, Fe 4; Ti 2	Rise at Ma due to Ca?
	23	K2	Fe 6; Fe 15	No. 23 the weaker line
	24	K2	Fe 2; Fe 3; Fe 10	No. 24 the strongest line
	25	K2	Fe 8	See No. 15. Rise at Ma unexplained
	26	?	Fe 2; Fe 5	
	27	K2	Fe 3; Fe 5	The stronger line. Responsible also for the maximum of line 2
	28	K2	V 2; Fe 4; V 2; Fe 3; V 2	Rise at Ma due to V
	29	(none)	Cr 5; Mn 5; Fe 4	Cr and Mn predominate. Cr (ultimate) line responsible for rise at Ma

OBSERVATIONAL MATERIAL

Atom	Note	Max.	Blends	Remarks
	30	F5	Ti 3; Fe 4	Maximum at F5 due to Fe+
	31	Ma	Fe 5; Ca 4	Ca produces rise at Ma
	32	Ma	Fe 2; Fe 2; Fe 2	Rise at Ma unexplained
	33	K2	Sr+	Due entirely to Sr+
	34	K2−K5	Fe 3; Fe? 3; V−Fe?3; Fe 5	
	35	?	Fe 4; Fe, −3; Mn 3; Co 3; −, 1; Fe−Cr 3	Too heavily blended
	36	Go	Co 4 d?; Fe 4; Ti 4	Maximum at Go unexplained. Rise at Ma due to Ti (ultimate line)
	37	K2	Cr, La, Mn, Ni, Fe 2; Ti, Fe 2; Al?; Fe 2	See Rowland, p. 37. An Fe+ line responsible
	38	K5	Fe 2; Fe 1; Ni 2	
	39	none	Fe 4; Fe, Mn 3; Nd?	Rise at Ma unexplained
	40	(Ma)	Fe 4; Ag 3	Rise at Ma probably due to Ag
	41	F5	Fe 4; Fe 3; −3; −3	Maximum certainly due to Fe+. Neutral Fe causes rise in cool classes
	42	F5	Fe 3; −3; −1; −1N	Maximum due to ionized iron
	43	F5	Mn 2; Fe 8	Maximum due to Fe+; later rise perhaps due to Mn
Zn	1	Go	Zn 3	Unblended
	2	Go	Zn 3	Unblended
Sr	1	none	Sr 1; Fe 4	Fe probably predominates, except perhaps at the lowest temperatures
	2	K2	Fe 2; Sr 5d?; Fe 3 d?	Fe probably strong, but Sr responsible for part of maximum at K2
	3	Ma	Fe, Zr 2; Fe 4; Fe 2; La, Y 1 Nd? Sr 8; Fe 4; Ti 3	Maximum uncertain owing to heavy blending
Y	1	Go	Cr 1; −1; Sc, Fe 3	Y+ gives the maximum
	2	F5?	Fe 3; −3	Maximum ill determined, but probably due to Y+
	3	Go?	−1	Remark in Rowland: — in zircon but not in Zr
Ba	1	none	Ba 8	Unblended

CONSISTENCY OF RESULTS

The preceding tabulation summarizes the present state of the observational material bearing on the positions of the maxima of absorption lines. The comparison with theory is an important and difficult problem. The theoretical formulae contain as variables the temperature and the pressure; and the fractional concentration, n_r, is very sensitive to changes in both these variables. It would therefore be possible to satisfy almost any observations by varying the two quantities jointly; but this procedure would furnish no useful test of the theory. The test made in the present chapter will involve the calculation of the temperature scale, with the partial electron pressure,

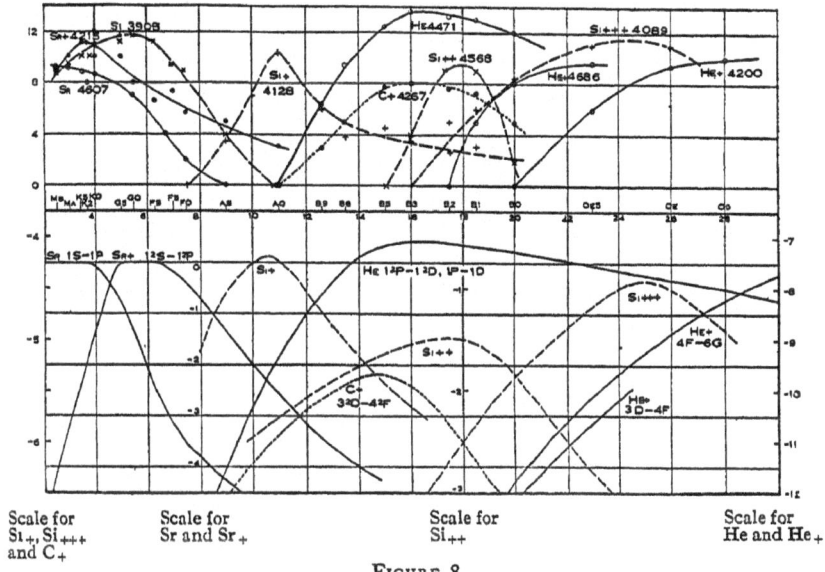

FIGURE 8

Reproduced from H.C.256, 1924. Comparison between observation and ionization theory for the hotter stars. The observations are contained in the upper part of the diagram, and the theoretical curves (based on a partial electron pressure 1.3×10^{-4} atmospheres) are given in the lower part of the figure. For the upper half, ordinates are the observed intensities contained in Table XIX; abscissae are spectral classes from the Draper Catalogue. In the lower part of the figure, ordinates are logarithms of computed fractional concentrations; abscissae are temperatures in thousands of degrees. The abscissae of the upper and lower diagrams have been adjusted so that the observed and computed maxima coincide, thus forming a preliminary temperature scale.

P_e, assumed constant. It is certain that this condition is not satisfied in practice, and a more rigorous treatment, which allows for the differences in partial electron pressure, is contained in the chapter that follows. But with the object of examining the consistency of the derived temperature scale, the present test is made under the assumption that the partial electron pressure is constant and equal to about 1.3×10^{-4} atmospheres.

The resulting scale of temperatures for the reversing layers of the corresponding classes is contained in the table that follows. Successive columns contain the element that is utilized, the spectral class at which its lines attain maximum, and the corresponding temperature derived from the equations of Chapter VII.

Element	Maximum	Temperature	Element	Maximum	Temperature
He+	O	35000°	Ti	K2–K5	3500°
Si+++	O	25000	Mn	K2	5000
Si++	B2–B1	18000	Fe	K2	5000
He	B3	16000	V	K5	3500
C+	B3	16000	Cr	K5	3500
Si+	A0	11000	Sr+	Ma	6000
H	A0	10000	Ba+	None	5500
*Zn	G0	8000	Ca	Ma	4500
*Ca+	K0	6000			

* Estimates by Menzel, H. C. 258, 1924.

CHAPTER IX

THE IONIZATION TEMPERATURE SCALE

A PRELIMINARY application of the observed maxima of absorption lines, in the formation of a stellar temperature scale, was given at the end of the preceding chapter. The temperatures were obtained on the assumption that P_e, the partial electron pressure in the reversing layer, was constant for all lines and equal to 1.3×10^{-4} atmospheres. Striking inconsistencies appear in this preliminary table of temperatures. As Menzel[1] has remarked, the maxima of most of the metallic arc lines occur in stars cooler than the ionization theory, on the stated assumptions, would predict. The ultimate lines of the ionized atoms of calcium, strontium, and barium show especially large inconsistencies. The temperatures of the maxima for these atoms, deduced from the ionization formula on the assumption that $P_e = 1.3 \times 10^{-4}$ atmospheres, are about 3000° higher than the measured temperatures of the classes at which the maxima occur, as deduced from the color indices.

The following suggestion has been advanced by Fowler and Milne[2] to account for the observed deviations of Ca+, Sr+, and Ba+. "For the maximum of the principal line of an ionized atom, the fraction of atoms in the required state is almost unity. . . . On the other hand . . . at the maxima of subordinate lines the fraction of atoms in the required state is from 10^{-3} to 10^{-5}. . . . Thus atoms in the required state are 10^3 to 10^5 times as abundant for intense principal lines as for intense subordinate lines. It follows that principal lines must originate at much higher levels in the stellar atmosphere than subordinate lines, and consequently at much smaller pressures."

[1] H. C. 258, 1924.   [2] M. N. R. A. S., 84, 499, 1924.

134 IONIZATION TEMPERATURE SCALE

It appears that the behavior of the ionized atoms of the alkaline earths can be satisfactorily explained in this way. The further suggestion was made that a similar effect might be expected for atoms of low excitation potential, such as manganese and magnesium.

The possibility of varying P_e as well as T in the formula for the theoretical maximum places the investigation on a rather different footing. Any temperature (within wide limits) may now be obtained for the theoretical maximum of a line by appropriately varying the partial pressure. The stellar temperature scale cannot, in such a case, be fixed merely from a knowledge of the critical potentials and the observed maxima, without introducing other considerations. It is necessary to find a way of determining the appropriate partial pressures.

The procedure that will here be followed consists essentially in a calibration and an extrapolation. The temperature scale from Ma to Fo is regarded as known from spectrophotometric data. Within this range, the theoretical and observed maxima are compared. The possibility of finding a value of P_e appropriate to a given atomic state is next examined. Finally, a method of estimating P_e will be justified for the cooler stars, within the limits of accuracy permitted by the data, and will be extended by simple extrapolation to the formation of a temperature scale for the hotter stars, where the temperatures cannot be safely estimated from the color indices.

The salient point is that complete absorption will occur for any line at a depth that is inversely proportional to the abundance of the corresponding state of the atom. No light in this wave-length reaches the exterior from any lower level, and the deepest level from which the line originates therefore forms a lower boundary to the effective portion of the atom in question. The " effective level " from which a line comes is probably best regarded as the level at which the effective atoms above the " lower boundary " have their median frequency. Clearly the partial pressure will differ at different effective levels, and thus abundance has a direct influence on the appropriate value of the partial pressure.

The theory with which we have so far been concerned deals with the *excited fraction* of the total amount of the element which is present. A knowledge of this quantity suffices for specifying the variation of intensity for the lines of any one element. But the absolute abundance of a given atomic state varies jointly with the fractional concentration of the appropriate state and the *total* amount of the element present. Now, for the first time, the absolute abundance of different atomic species becomes of possible importance, as a factor affecting the depth from which radiation corresponding to the given atom will penetrate. Fowler and Milne [3] rightly claimed that their method of maxima eliminated questions of relative abundance, "if P_e can be regarded as known . . . [and constant]. The proper value of P_e must be a function of the abundance of the atom in question relative to free electrons."

The question of relative abundances of elements in the reversing layer is discussed [4] in Chapter XIII. It may be mentioned that the abundances there deduced depend upon estimates of *marginal appearance*. Probably all lines are unsaturated at marginal appearance, that is, there are not enough suitable atoms present completely to absorb *all* the incident light of the appropriate wave-length. Hence all suitable atoms present, as far down as the photosphere, where general opacity begins to render the gas hazy, are actually contributing to the line. At marginal appearance, then, all the intensity phenomena are probably due to pure abundance, and considerations of level are eliminated. The deduced abundances are therefore independent of effects such as are discussed in the present chapter, and the results of Chapter XIII may be cited as giving evidence that the stellar abundances, for all the atoms here to be considered except barium, have a range with only a factor of ten, which is negligible in comparison with the quan-

[3] M. N. R. A. S., **84**, 499, 1924.   [4] Chapter XIII, p. 177.

tities to be discussed. The relative abundance of different atomic species will therefore be neglected in what follows, although, with more accurate data than are now available, it should become a factor of importance.

Fractional concentrations, as derived from the ionization formula, govern the effective level at which absorption takes place. Fowler and Milne, as was pointed out earlier, suggested that the higher the fractional concentration at maximum, the higher the level and the lower the partial pressure from which the line originates. They suggested that the pressure for a principal line at maximum is from 10^{-3} to 10^{-5} of the corresponding value for a subordinate line.

The assumption now introduced is, in effect, that the absorbing efficiency of individual atoms is the same. The partial pressure at the level from which a line originates should then vary inversely as the fractional concentration at maximum. In other words, the product

$$P_e \times n_{r(max)}$$

should be constant, when P_e is deduced from the class at which the *observed maximum* occurs.

The quantity $n_{r(max)}$ depends primarily on the excitation potential, and varies but slowly with T_{max}. It is given by the expression*

$$n_{r(max)} = \frac{\chi_1^{(r)} + \tfrac{5}{2}kT_{max}}{\chi_1 + \tfrac{5}{2}kT_{max}} \cdot \frac{q_1^{(r)} e^{-(\chi_1 - \chi_1^{(r)})/kT_{max}}}{b_1(T_{max})}$$

For subordinate lines, P_e is given by the expression

$$\frac{\chi_1^{(r)} + \tfrac{5}{2}kT}{\chi_1 - \chi_1^{(r)}} \cdot \frac{(2\pi m)^{\tfrac{3}{2}}(kT)^{\tfrac{5}{2}} \sigma_1 e^{-\chi_1/kT}}{h^3 b_1(T)}$$

and this quantity is extremely sensitive to change in T_{max}.

For ultimate lines, where the excitation potential is equal to zero, and $n_{r(max)}$ accordingly reduces to unity, the value of P_e should be *equal* to the constant product predicted in a previous

* For notation, see Chapter VII, p. 106.

paragraph. Fowler and Milne suggested a partial electron pressure of 10^{-7} to 10^{-8} for Ca+ on the basis of a maximum at K0, assumed temperature 4500°. This is the effective temperature of the class, deduced spectrophotometrically, and "the reversing layer should be at a lower temperature — its average temperature should be in the neighborhood of, or somewhat lower than, the Schwarzschild boundary temperature,** which is some 15-20 per cent lower than the effective temperature." The value 4000° is therefore adopted here for K0. For this value P_e becomes 3.24×10^{-9} for Ca+; for Sr+ (K2,3500°), $P_e = 4.6 \times 10^{-10}$, and for Ba+ (Ma? 3000°), $P_e = 6 \times 10^{-11}$. The maximum for Sr+ is the best determined of the three, as the Ca+ lines are too strong and too far into the violet for an accurate estimate among the cooler stars, and the Ba+ line is rather faint, and is heavily blended. The constant product may then be expected to be of the order of 10^{-10}.

The prediction is examined in the table that follows. The temperature of the class at which the lines attain maximum is assumed from spectrophotometric data, and is expressed to the

TABLE XX

Atom	Ionization Potential	Excitation Potential	Max.	T_{max}	$\log P_e$	$\log n_r$	Sum
Mg+	14.97	8.83	A3	9000°	$\bar{5}.99$	$\bar{5}.32$	$\overline{10}.3$
Ca	6.09	1.88	Ma	3000	$\bar{8}.63$	$\bar{3}.64$	$\overline{10}.3$
Ti	6.5	0.84	K2	3500	$\bar{7}.8$	$\bar{2}.7$	$\bar{8}.5$
Cr	6.75	0.94	K5	3000	$\bar{9}.75$	$\bar{2}.39$	$\overline{10}.1$
Mn	7.41	2.16	K2	3500	$\bar{8}.69$	$\bar{3}.77$	$\overline{10}.4$
Zn	9.35	4.01	G0	5600	$\bar{7}.8$	$\bar{3}.0$	$\overline{10}.8$
Ca+	11.82	0.00	K0	4000	$\bar{9}.46$	0.00	$\bar{9}.5$
Sr+	10.98	0.00	K2	3500	$\overline{10}.60$	0.00	$\overline{10}.6$
Ba+	9.96	0.00	Ma?	3000	$\overline{11}.89$	0.00	$\overline{11}.9$
Mg	7.61	2.67	K0?	4000	$\bar{7}.25$	$\bar{3}.25$	$\overline{10}.5$

** The Schwarzschild approximation to the boundary temperature is given by the expression

$$T_1{}^4 = \tfrac{1}{2}T_0{}^4$$

where T_0 is the effective temperature and T_1 the boundary temperature.

nearest five hundred degrees. Successive columns give the atom, the critical potentials in volts, the spectral class at which maximum occurs, the assumed T_{max}, log P_e calculated from the theory, log $n_{r(max)}$, and the sum of the quantities in the two preceding columns. The only quantity that is not fixed by the laboratory data is T_{max}, which is derived from the data presented in Chapter II. It will be seen that the quantity entered in the last column is sensibly constant, and equal to about -10, in accordance with prediction. All available maxima have been used.

It appears that the foregoing evidence constitutes a fair and satisfactory test of the Fowler-Milne equations, and that, in the region in which the test can be applied, the agreement with theory is as close as can be expected from the material. It also appears that the "serious and undoubtedly real" discordance of theory and observation, quoted by Menzel in the discussion of the maxima observed by him, is removed by introducing these considerations of level.

When the theory has been applied and justified for the classes where the temperature scale is well determined by other methods, it may be extrapolated to fix the temperature scale for the hotter stars. As before, the fractional concentration at maximum varies but slowly with T, and T_{max} is determined mainly by P_e. If now P_e be so chosen that $P_e \times n_{r(max)}$ is always approximately equal to 10^{-10}, the value of T derived from the equations will be the appropriate one for the class in question. This value of T has to be found by trial. It so happens that the temperatures thus obtained are not very different from those originally predicted without entering into considerations of effective level. The excitation potentials of the highly ionized stages of the lighter elements are invariably large, and all lead to values of P_e of the order of 10^{-4}. It is to be noted that values of P_e *greater* than 10^{-4} are not indicated.

The following tabulation represents the resulting temperature scale for the hotter stars. It must be remembered that T_{max} is

TEMPERATURES FOR DRAPER CLASSES 139

TABLE XXI

Atom	Ionization Potential	Excitation Potential	Max.	T_{max}
He+	54.2	48.2	O	35000°
C+	24.3	18.0	B3	16000
He	24.7	21.1	B3	16000
Si++	31.7	4.8	B2−B1	18000
Si+++	45.0	24.0	O	25000

here the *derived* quantity, whereas in Table XX it was the known quantity used for calibration.

The values given in the preceding table constitute the only contribution that can be made by this form of ionization theory to the formation of a stellar temperature scale. Values assigned to intermediate classes must be conjectural. From the observed changes of intensity from class to class, temperatures may be interpolated roughly, and a temperature scale, formed on these general grounds, is reproduced in Table XXII. Values not derived from observed maxima are italicized.

TABLE XXII

Class	Temperature	Class	Temperature
Ma	3000°	A3	9000°
K5	3000	A0	10000
K2	3500	B8	13500
K0	4000	B5	*15000*
G5	*5000*	B3	17000
G0	5600	B1.5	18000
F5	*7000*	B0	20000
F0	*7500*	O	25000
A5	*8400*	to	35000

CHAPTER X

EFFECTS OF ABSOLUTE MAGNITUDE UPON THE SPECTRUM

DIFFERENCES between the spectra of stars of the same spectral class have long been recognized. The empirical correlation of relative line intensities with absolute magnitude was made the basis for the estimation of spectroscopic parallaxes.[1] Such differences within a class were later related in a qualitative way to differences of pressure, in conjunction with the theory of thermal ionization, and have been regarded as corroborative evidence that the type of process contemplated by that theory actually represents what goes on in the atmospheres of the stars.

In the present chapter the theory of the various effects will first be discussed, and later the predictions from the theory will be compared with observational data.

INFLUENCE OF SURFACE GRAVITY ON THE SPECTRUM

The first theoretical discussion of the effects of absolute magnitude upon the stellar spectrum seems to have been made by Pannekoek,[2] who pointed out that "stars of the same spectral class . . . will show differences depending solely on . . . g/k," where g is the surface gravity, and k the absorption coefficient. Pannekoek considered all stars of the same spectral class to have the same temperature, and for the purposes of his argument the differences in temperature between giants and dwarfs can be neglected, although actually they may for other reasons have a noticeable effect on the spectrum. If k be regarded as constant, a plausible assumption for various reasons,[3] "the physical quantity, directly given by the spectra used for the deter-

---

[1] Adams and Kohlschütter, Mt. W. Contr. 89, 1914.
[2] B. A. N. 19, 1922.
[3] Milne, Phil. Mag., 47, 209, 1924.

mination of spectroscopic parallaxes is the gravitation at the surface of the star." [4] The relation between the surface gravity and the pressure is given by

$$p = (g/k)\tau$$

where τ is the "homogeneous depth." The pressure is then directly proportional to the surface gravity.

INFLUENCE OF PRESSURE ON THE SPECTRUM

Lowered pressure increases the *degree of ionization*. The tendency of the atoms to lose electrons by thermal ionization should depend solely on their energy supply, and should thus be independent of the pressure. The total absorbing power of the gas will, however, depend on the *number* of suitable atoms that it contains, not upon their *rate* of formation. The number of suitable ionized atoms present at any moment in the atmosphere is a function not only of the rate at which ionization proceeds, but also of the rate of recombination. The more readily recombination takes place, the larger is the number of effective neutral atoms, and the smaller the number of effective ionized atoms, when a steady state is attained. The rate of recombination, which depends upon the probability of a suitable encounter between an ionized atom and a free electron, will increase with the pressure — more accurately, with the partial pressure of free electrons.

The higher the pressure, therefore, the greater the number of neutral atoms, and the smaller the number of ionized atoms. This argument explains at once the strength of the neutral (arc) lines in the spectra of stars of low luminosity (high surface gravity), and the predominance of ionized (spark) lines for absolutely bright stars (low surface gravity, resulting chiefly from large radius). Low surface gravity, then, increases the number of ionized atoms present by discouraging recombination.

It should be noted that any tendency to extensive ionization will increase the concentration of free electrons and tend to en-

[4] Pannekoek, B. A. N. 19, 1922.

courage recombination, thus counteracting the effect of low surface gravity. The effect of an increased concentration of free electrons will not, however, attain the magnitude of the surface gravity effect, since even for the hottest stars examined, three electrons appear to be the largest number that can be thermally removed under reversing layer conditions.

The theoretical effect of lowering the pressure has been discussed by Stewart,[5] who, after alluding to the importance of the surface gravity, suggested that the ultimate lines of neutral atoms easier to ionize than the average should be weakened by low pressure, and that the corresponding enhanced lines should be strengthened. For atoms harder to ionize than the average the reverse should be the case for the two classes of lines. From this standpoint he showed that the absolute magnitude effects might be qualitatively accounted for. The "average ionization potential" was the average for the lines used in the estimates; Stewart adopted the value of six volts for Classes F to K.

Effect of Temperature and Density Gradients upon the Spectrum of a Star

There is another respect, recently analyzed by Stewart,[6] in which the spectrum of a giant may be expected to differ from that of the corresponding dwarf. He points out that "in a giant, owing to the small density, there is more material overlying the photosphere than in a dwarf having the same effective temperature; while at the same time the density in the photospheric region is less in the giant, owing to the low gravity." These conditions furnish an interpretation of the increased blackness and sharpness of the lines in giant stars, as compared with the corresponding dwarfs. The absorption lines in giants are *blacker* because there is more matter above the photosphere than in dwarfs; they are *sharper* because the effective level at which the lines originate is at a lower pressure in the giant than in the dwarf, owing to the smaller pressure gradient in the giant

---

[5] Pop. Ast., 31, 88, 1923.   [6] Pop. Ast., in press.

star, and to its lower surface gravity. The difference in line quality between a giant and a dwarf is at once obvious from the spectra, and this effect renders direct comparisons of estimated line-intensities a matter of extreme difficulty. It is an effect that must be taken into account in examining the agreement between the observations and the theory.

Stewart's argument also suggests the answer to an important question raised by Pannekoek [7] in the course of his discussion of the absolute magnitude effect. The latter remarks that "the general decrease of luminosity with advancing type for the same value of relative line-intensity, which is shown . . . by most reduction curves . . . corresponds to the decrease in σ, as for the same g and smaller R smaller surface brightness means smaller luminosity. If we take account, however, of the direct influence of temperature on ionization, which acts much more strongly in the opposite direction, we must expect equal ionization in the more advanced types for much smaller g and higher luminosities, contrary to the empirical reduction curves. It looks as if this effect is compensated by some other direct influence of temperature on the spectrum." [8]

The influence suspected by Pannekoek may be found, at least in part, in the "theoretical decrease with increasing temperature and density in the quantity of material overlying the photosphere. Thus the contrast between line and continuous background tends to become less along the giant series M–B (since, furthermore, for the same abundance of active material, a given line is formed always at the same depth)." [9] This suggestion was advanced by Stewart to account for the observed displacement, towards cooler classes, of the maxima of absorption lines discussed in Chapter X. It is certain that some such factor will be operative in the reversing layer, but it is believed that the burden of the shift of maxima should be borne by the effective

---

[7] B. A. N. 19, 1922.

[8] In Pannekoek's notation, σ is surface brightness, R is radius, and g, surface gravity.

[9] Stewart, Pop. Ast., in press.

level, which has been discussed in more detail in the preceding chapter. It would be of interest to compare the two effects quantitatively, but the effect of temperature gradient has not yet formed the basis of numerical predictions.

PREDICTED EFFECTS ON INDIVIDUAL LINES

The discussion involving the average ionization potential appears to permit of more rigorous treatment. Suppose the "average ionization potential" of Stewart's discussion to be replaced by the ionization potential corresponding to the atoms whose lines are at maximum for the class in question. It then follows directly from theory that the effects of lowered pressure on the different classes of lines will be as below:

| Atom | Line | Effect of lowered pressure | |
		Hotter than class for maximum	Cooler than class for maximum
Neutral	Ultimate	Weakened	....
Neutral	Subordinate	Weakened	Weakened
Ionized	Ultimate	Weakened	Strengthened
Ionized	Subordinate	Weakened	Strengthened

It is especially to be noted that all lines should theoretically be weakened in passing from dwarf to giant, excepting the lines of an ionized atom at temperatures lower than those required to bring them to maximum. This leaves out of account the effect of photospheric depth, which will be introduced later as a correcting factor.

The case of the ultimate lines of the ionized atoms is of especial interest. At their maximum, if the Fowler-Milne theory is correct, ionization is almost complete, and more than 99 per cent of the element is giving the ionized ultimate lines. At a temperature higher than that required for maximum, lowered pressure can "increase" the ionization only by the removal of the second electron. By this process the intensity of the ionized ultimate lines is *decreased*, since the number of singly ionized atoms is thereby reduced. The fall from maximum towards the hotter stars, which is displayed by the ionized lines of Ca+, Sr+, and Ba+ can be due only to the progress of second ionization, and there seems to be no escape from the conclusion that the ultimate lines of the ionized atom should theoretically decrease in strength, with lowered pressure, for stars hotter than those required to bring the lines to maximum. The point is made increasingly clear when it is recalled that, at the maximum, all of the substance is presumably at work giving the lines in question. It is not therefore possible to increase the number of active atoms by any process whatever that involves merely a change in pressure.

For ionized subordinate lines the theoretical effect should be the same as for the ultimate lines, for the fall after maximum is here again caused by the increase in the number of doubly ionized atoms, and the consequent decrease in the number of those singly ionized. Thus, although the subordinate lines are not already using all the available atoms at maximum, so that increased intensity with lowered pressure is possible, it would still appear that they should be weakened at temperatures higher than that corresponding to maximum intensity in the spectral sequence.

The pure pressure effects just discussed will be superposed upon the Stewart effect, which depends upon the photospheric depth. The latter will cause a general increase in the strength of all lines from dwarf to giant, as a result of the greater amount of matter lying above the photosphere in the giant. The two effects are observed together when direct intensity measures are employed, such as the estimates embodied in Chapter VIII, while the pressure effect is given almost purely when differential estimates of intensity for the same spectrum are used, as in most investigations of spectroscopic parallax. The observational evidence from both sources will now be put forward, in order to examine the sufficiency of the theories that have been advanced to account for the absolute magnitude effects.

The empirical relations used in the estimation of spectroscopic parallax should provide material for examining the simple pressure effect, as they are derived from the ratio of two lines in the same spectrum. Unfortunately the line ratios actually in use were selected because they were convenient to measure, and gave (empirically) consistent results, not for reasons of theoretical tractability. Fourteen line ratios are used, for example, by Harper and Young,[10] but only four of these consist of pairs of unblended lines with known series relations. It is only for such lines that a useful test of theory can be made.

TABLE XXIII

$\frac{4071}{4077}$	F0	F5	G0	G5	K0	K5	M0	M5
M = +7	+ 8.8	+ 9.2	+11.0	+13.2				
+6					+13.3			
+5	+ 3.5	+ 5.0	+ 7.4	+ 9.6				
+4					+10.0	+10.8		
+3	− 1.8	+ 0.5	+ 3.8	+ 6.7				
+2					+ 6.7	+ 7.0	+7.6	+10.0
+1	− 7.2	− 3.8	+ 0.2	+ 3.2				
0					+ 3.5	+ 3.4	+3.2	+ 3.0
−1	−12.9	− 8.2	− 3.2	− 0.3				
−2					+ 0.3	− 0.2	−1.4	− 3.7
−3		−12.1	− 7.3	− 3.8				
−4					− 2.8	− 4.2	−5.8	−10.0

$\frac{4215}{4250}$	F0	F5	G0	G5	K0	K5		
M = +6					− 2.0	− 1.4		
+5	− 1.5	− 0.5	0.0	+ 1.0				
+4					+ 1.2	+ 1.3		
+3	+ 1.8	+ 3.3	+ 4.4	+ 4.5				
+2					+ 5.0	+ 3.0		
+1	+ 4.8	+ 6.5	+ 7.8	+ 7.6				
0					+ 6.5	+ 4.7		
−1	+ 8.6	+ 9.3	+11.0	+11.0				
−2	+12.0	+ 8.6	+ 8.6	+ 8.2				

$\frac{4247}{4250}$	F0	F5	G0	G5	K0	$\frac{4455}{4494}$	G5	K0	K5
M = +6					−18.1	M = +6		+7.4	+6.2
+5	−3.7	−5.8	−9.6	−14.7		+5	+6.2		
+4					−15.3	+4		+4.3	+3.0
+3	−1.0	−3.3	−6.6	−11.0		+3	+3.0		
+2					−12.4	+2		+1.3	0.0
+1	+1.7	−0.6	−3.7	− 7.5		+1	0.0		
0					− 9.4	0		0.0	−2.0
−1	+4.5	+2.0	−0.9	− 4.0		−1	+2.6		
−2					− 6.6	−2		+2.0	+0.7
−3	+7.2	+4.7	+2.0	− 0.2		−3	+5.2		

[10] Pub. Dom. Ap. Obs., 3, 1, 1924.

OBSERVATIONAL DATA 147

The preceding table contains a transcription of the reduction-curve material given by Harper and Young for the four pairs of lines mentioned. Tabulated quantities are the "step differences" for the classes at the heads of the columns, and the absolute magnitudes contained in the first column.

Presumably the irregularities of the observed curves have been smoothed out in forming the reduction table, but the figures will certainly give an indication of the *direction* in which a given line is affected by absolute magnitude.

The predicted effect of lowered pressure upon the lines involved is contained in the table that follows:

Line	Source	Max.	Effect of lowered pressure
4071	Fe (sub)	K2	weakened throughout
4077	Sr+ (ult)	K2	strengthened in M, weakened in G and F
4215	Sr+ (ult)	K2	strengthened in M, weakened in G and F
4247	Sc+ (?)	F0	strengthened throughout range
4250	Fe (sub)	K2	weakened throughout
4455	Ca (sub)	Ma	weakened throughout
4494	Fe (sub)	K2	weakened throughout

The predicted changes in the line ratios with lowered pressure are therefore as follows:

$\frac{4071}{4077}$ increased from F0 to K0, decreased from M0 to M5

$\frac{4215}{4250}$ increased from F0 to K0, decreased from M0 to M5

$\frac{4247}{4250}$ increased throughout

$\frac{4455}{4494}$ indeterminate

The ratio $\frac{4247}{4250}$ behaves in exact accordance with prediction, and $\frac{4455}{4494}$, which decreases and then increases again, offers no evidence for or against the theory. The two remaining ratios, involving the two Sr+ lines, display a lack of agreement with theory for the F and G classes, apparently owing to the strengthening of the Sr+ lines with high luminosity, even at temperatures higher than those at which they attain maximum intensity. The strengthening of Sr+ with high luminosity is one of the best-attested facts of observational astrophysics, and it is a serious deficiency in theory if the observed behavior of the lines in the hotter stars cannot be explained. The question will be further discussed presently.

The material obtained by the writer, and summarized in a preceding chapter,[11] may be used in making a test of the predicted pressure effects by means of direct estimates. As was pointed out above, the lines of a giant are stronger than those of a dwarf, owing to the greater photospheric depth in the former. The practical difficulty of making comparable estimates upon sharp and somewhat hazy lines must also be considered in the discussion of the results. Clearly some numerical correction is required, in order to allow for the Stewart effect, and this has been done in a somewhat arbitrary manner in forming Table XXIV. It is assumed that the mean increase in intensity for such lines as are strengthened will be equal to the mean decrease in intensity for such lines as are weakened. For each spectral class this assumption provides a correcting factor, which never exceeds one scale unit.

The table that follows contains the material derived from the measures enumerated in Chapter VIII, and from other sources, bearing on the intensity differences between giants and dwarfs of the same spectral class. All the available estimates have been used. Successive columns give the line, the atom, the predicted behavior, and the observed difference in the sense giant-dwarf, for the Classes F0, F2, F5, F8, G0, and G5. The symbols u, n, and *, following the atom, denote ultimate, neutral, and enhanced lines, respectively. The number of stars contributing to each entry will be seen from the list on p. 119, Chapter VIII. The notation is as follows: 0=no change; ± 0 = between 0 and 1; ± 1= between 1 and 2; ± 2= between 2 and 3; and so on. The values for K0 are taken from Menzel's measures [12] of ε Indi and α Tauri, " the scale of intensities being (0) no difference, (1) a

[11] P. 121.   [12] H. C. 258, 1924.

OBSERVATIONAL DATA

little stronger . . ., (2) much stronger, (3) very much stronger." The signs from Menzel's table are reversed, in accordance with the notation used in the present table. In the column headed Ma are the signs indicating the direction in which the corresponding lines are affected in that class,[13] for which quantitative measures have not been published. The letters " s " and " w " in the column headed Go refer to strengthening or weakening of lines, as observed by Baxandall[14] in a comparison of the solar spectrum with that of Capella. Baxandall's estimates are inserted to supplement the present material. The numerous gaps in the table result from the difficulty of seeing the fainter lines in the dwarf spectrum.

TABLE XXIV

Line	Element	Predicted Effect			Observed Effect						
		−	o	+	Fo	F2	F5	Go	G5	Ko	Ma
3933	Ca+ *	Fo–G5	Ko	Ma	+1	..	..	..	..	o?	+
3944	Al u	Fo–Ma	..	..	+o	+	−2	..	..	..	−
3953	Fe n	Fo–Ko	(K2)	..	+o	+	−2	..	..	..	..
3961	Al u	Fo–Ma	..	..	+o	+	−2	..	..	..	−
3968	Ca+ *	Fo–G5	Ko	Ma	+1	..	−1	..	..	o?	+
3999	Ti u	Fo–Ma	..	..	+o	..	−1	.	..	..	..
4005	Fe n	Fo–Ma	..	..	+o	..	o	..	..	..	..
4031	Mn u	Fo–Ma	..	..	−2	−1	−1	..	..	o	..
4041	Mn n	Fo–Ma	..	..	−1	−1	−2	..	..	−o	..
4046	Fe n	Fo–Ma	..	..	−1	−1	o	−1	−1	−2	..
4064	Fe n	Fo–Ma	..	..	−1	−2	..	+2	o	−2	..
4068	FeMn n	Fo–Ma	..	..	−1	..	..	+1	−1	..	..
4072	Fe, − n	Fo–Ma	..	..	o	+4	..	..	−1	−2	..
4077	Sr+ *	Fo–Ko	K2	K5–Ma	o	o	+1	s	o	+3	+
4084	Fe n	Fo–Ko	..	..	−1	+1	o	o	..	..	..
4101	H n	Fo–Ma	..	..	−1	−1	+1	−4	o	−1	+
4132	Fe n	Fo–Ma	..	..	o	+2	+1	..	+o	−1	..
4135	Fe n	Fo–Ma	..	..	−1	..	o	..	+o	..	..
4144	Fe n	Fo–Ma	..	..	−1	o	+2	o	+o	−1	..
4167	?				−1	..	−1	−1w	o	..	..

[13] Adams, Pub. A. S. P., 28, 278, 1916; Adams and Joy, Pub. A. S. P., 36, 142, 1924.
[14] Pub. Solar Phys. Com., 1910.

Line	Element	Predicted Effect			Observed Effect						
		−	o	+	Fo	F2	F5	Go	G5	Ko	Ma
4172	Fe+ *	..	Fo	F2–Ma	+1	+1	+3	+1	+0	..	..
4177	Fe+ *	..	Fo	F2–Ma	+1	..	+2	..	..	..	..
4215	Sr+ *	Fo–Ko	K2	K5–Ma	0	0	+2	+3	+0	+1	+
4227	Ca u	Fo–Ma	..	..	−1	−1	0	−2	−0	−2	−
4247	Sc+ *	..	Fo	F2–Ma	0	..	+1	..	+1	..	..
4250	Fe n	Fo–Ma	..	..	0	−2	−1	−2	−1	−2	..
4254	Cr u	Fo–Ma	..	..	−1	..	−1	−1	−1	−1	..
4260	Fe n	Fo–Ma	..	..	−1	−2	−1	−2	..	−1	−
4272	Fe n	Fo–Ma	..	..	−1	−2	..	0	+0	−1	..
4275	Br u	Fo–Ma	..	..	−1	..	..	0	+0	−1	..
4290	Cr u	Fo–Ma	..	..	0	−2	0	..	+0	−1	..
4298	Ti, Ca	Fo–Ma	..	..	−1	−2	0	..	0	..	..
4308	Fe n	Fo–Ma	..	..	0	+1	0	..	..	−1	..
4315	Fe+ *	..	Fo	F2–Ma	0	+1	+2	..	..	..	..
4321	Sc+ *	..	Fo	F2–Ma	+1	..	..	s	..	..	..
4326	Fe n	Fo–Ma	..	..	0	+1	+1	−2w	0	−2	..
4340	H n	Fo–Ma	..	..	−2	−7	−1	−4	−0	−1	+
4352	CrMg	Fo–Ma?	..	..	0	+2	0	+1	+0	−3	..
4360	Cr n	Fo–Ma	..	..	0	..	+1	..	..	..	..
4370	Fe n	Fo–Ma	..	..	+1	..	0	..	+0	..	..
4376	Y+ *	..	Fo	F2–Ma	−1	+2	0	..	+1	+2	..
4383	Fe n	Fo–Ma	..	..	0	+3	+1	−2	−4	−2	..
4405	Fe n	Fo–Ma	..	..	−1	..	..	0	−1	−2	..
4415	Fe+ *	..	Fo	F2–Ma	+1	0	+3	0	0	−1	..
4435	Ca n	Fo–Ma	..	..	−1	..	+1	..	0	−2	−
4444	Ti+ *	..	Fo	F2–Ma	0	0	+1	s	−3	..	..
4455	Ca n	Fo–Ma	..	..	−2	..	−1	w	−1	−3	−
4476	Fe n	Fo–Ma	..	..	−1	..	+1	−1	..	..	..
4481	Mg+ *	..	..	Fo–Ma	−1	0	0	0	+0	+1	..
4490	Fe n	Fo–Ma	..	..	0	..	..	+1	−0	+2	..

It is seen from the table that the general agreement with the anticipations of theory is satisfactory, and that the deviations, when they occur, rarely exceed one unit. The agreement is not less good than would be expected of the material, since the measures are here used differentially. The majority of the discrepancies are apparently accidental; for example, the deviations shown by the first six entries in the first column are almost certainly the result of better definition in the giant spectrum.

There remains, however, the same discrepancy for the lines of Sr+ that was noted in the earlier part of the chapter. There can be no doubt that these lines are stronger in giants than in dwarfs.

The strengthening of the ionized lines of the alkaline earths is explained, when the spectra are examined, by the fact that the *neutral* lines are still fairly strong long after the ionized lines have passed their maximum — neutral strontium [15] is found at Fo and neutral calcium [16] at Ao. The lowered pressure, then, must increase the concentration of singly ionized atoms at the expense of the residual neutral atoms. There is, however, apparently no satisfactory theoretical explanation of the survival of large quantities of neutral calcium long after the ionized atoms have passed their maximum. The effects predicted above would appear to be the only ones that can be anticipated if the theory holds rigidly. Clearly some factor such as effective level must be further considered.

THE STRONTIUM LINES

The strontium problem is perhaps one that will lead to more comprehensible results when it is treated as a whole. It is impossible to resist the feeling that there is some definite abnormality associated with strontium. The "strontium stars" in the still earlier classes, where the lines 4215, 4077 appear with great intensity, and the Fo stars α Circini and γ Equulei, as well as the apparently erroneous absolute magnitudes obtained by the spectroscopic method for several other stars of low intrinsic luminosity, all point in some such direction.

It may be that these phenomena are a result of an abnormal abundance or distribution of the element. It is not, therefore, entirely necessary to assume that the theory is here at fault, although until the behavior of strontium has been satisfactorily interpreted, that possibility cannot be rejected. It is significant that calcium and barium show similar absolute magnitude behavior. In any case, the ionized strontium lines cannot be cited,

---

[15] Chapter V, p. 81.   [16] Chapter, V, p. 70.

as has sometimes been done, in demonstrating that the absolute magnitude effect is due to pressure. What is actually shown is that the concentration of singly ionized atoms is more greatly increased at the expense of the neutral atoms than it is reduced by the formation of doubly ionized atoms. Since a pressure effect operates by the discouragement of recombination, it would be inferred that the recombination of singly ionized atoms with electrons to form neutral atoms is less readily encouraged than the recombination of doubly ionized atoms with electrons to form singly ionized atoms. Evidently the problem is a complex one. If the maximum of the strontium lines were at F_5 (where theory first predicted it, and where the earlier measures actually placed it) there would be no anomaly to explain; but two independent observers [17] place it definitely at K_2 or K_0, and there can be little doubt that this is actually the correct position of the maximum.

The result of the study of absolute magnitude effects is disappointing. It appears that the observed phenomena are qualitatively explained in a satisfactory manner, as due to lowered pressure, or, more accurately, to low surface gravity. There is, however, a serious discrepancy in the case of the lines whose variation with absolute magnitude is perhaps best established, and upon which the most important results have been based. The results, being empirical, are of course unimpaired, and it would seem that the theory requires to be amended. Furthermore, it does not yet appear to be possible to use the observed changes of intensity for the direct estimation of pressure differences, because of the large number of variables involved and particularly because of the superposition of the pure pressure effect upon the effect of photospheric depth.

[17] Menzel, H. C. 258, 1924; Chapter VIII, p. 126.

PART III
ADDITIONAL DEDUCTIONS FROM IONIZATION THEORY

CHAPTER XI

THE ASTROPHYSICAL EVALUATION OF PHYSICAL CONSTANTS

In the opening chapter the statement was made that " the astrophysicist is obliged to assume [the validity of physical laws] in applying them to stellar conditions." The astrophysical evaluation of physical constants might therefore seem, from our avowed premises, to involve a circular argument. In certain special cases, however, the process appears to be legitimate, and the results of three investigations are contained in the present chapter. The first of these investigations involves the derivation of spectroscopic constants, assuming the series formula; the second consists of an extrapolation of the results of Chapter X to the estimation of unknown ionization potentials; and the third constitutes a discussion made possible by the knowledge of the stellar atmosphere that has been attained with the aid of ionization theory.

THE RYDBERG CONSTANT FOR HELIUM

The wave-lengths of a series of lines can be measured in the spectrum of a star, and the series identified with a series observed in the laboratory. The occurrence in stellar spectra of series that can be identified with the series given by terrestrial atoms presumably shows that similar relations govern the atomic processes in the two sources. That series formulae of the same type are applicable to the stellar and terrestrial atom is indeed rather an observational fact than an assumption. By inserting into the appropriate series formula the observed stellar frequencies, a physical constant involved may be evaluated, and the extent of the agreement with the corresponding value from the laboratory may be determined.

H. H. Plaskett[1] has measured the wave-lengths of the lines of the Pickering series (4F−mG) of He+ in the spectra of three O stars, incidentally separating the alternate Pickering lines from the Balmer lines for the first time. The formula that connects the frequencies of the lines with the constants associated with the atom is

$$\nu = N(E/e)^2 \left(\frac{1}{n^2} - \frac{1}{m^2} \right)$$

ν = frequency
N = the Rydberg Constant
E = mass of atom
e = mass of electron
$n = 3$
$m = 4, 5, 6. \ldots$

Plaskett discussed the theory, and derived from the measured wave-lengths of five lines the mean value of 109722.3 ±0.44 for the constant N. The value determined in the laboratory by Paschen is 109722.14 ±0.04. Plaskett's comment on the agreement is as follows: " It was not to be expected that there would be any startling changes. . . . It is of interest, however, to note that these " stellar " determinations *are* in agreement with the terrestrial values, in so far as it shows that *the implicit assumption of identical atomic structure, identical electrons, and identical laws of radiation on the earth and in the stars, is in some measure justified.*"

Critical Potentials

The theory outlined in the preceding chapters was used in determining the astrophysical behavior of lines corresponding to known series relations. When the validity of the theory has been established, it is possible, as was pointed out by the writer,[2] by Fowler and Milne,[3] and by Menzel,[4] to deduce the ionization potentials of lines of unknown series relations from their astrophysical behavior. The ionization potentials were estimated in this way for the table in Chapter I.

In general the observations show that the higher the ioniza-

---

[1] H. H. Plaskett, Pub. Dom. Ap. Obs., 2, 325, 1922.
[2] H. C. 256, 1924.
[3] M. N. R. A. S., 84, 499, 1924.
[4] H. C. 258, 1924.

tion potential, the higher the temperature at which the corresponding lines attain maximum. This is in strict accordance with theory. It is not possible to predict the exact form of the relation between temperature of maximum and ionization potential. For the observed cases in which $T_{max}=0$ (the ultimate lines), $\chi_1-\chi_1^{(r)}=0$. It would appear that T_{max} should approach zero as χ_1 approaches zero. But in this case $\chi_1^{(r)}$, (the negative energy of the excited state, which must always be less than χ_1) also approaches zero, and the relation becomes indeterminate. The form of the curve as χ_1 approaches zero has merely a theoretical interest, as no known element has an ionization potential of less than four volts. In the present application the relation will be treated as an empirical one. The curves given by the writer and by Menzel for the relation between ionization potential and T_{max} display a good general regularity, and the deviations, as was pointed out in a previous chapter,[5] probably arise from differences of effective level. Owing to this source of irregularity, great accuracy is not to be anticipated in the deduced ionization potentials. The effective level is at the greatest height for lines of low excitation potential. The excitation potentials corresponding to the astrophysically important lines of the once, twice, and thrice ionized atoms in the hotter stars are in all known cases high, and thus the error introduced by neglecting to correct for effective level is small. The error introduced by an excitation potential of the wrong order is, moreover, a constant and not a percentage error, and thus becomes less serious in estimating high ionization potentials. Accordingly the deduced ionization potentials will probably be of the right order.

The relation connecting ionization potential and T_{max} may, for our purposes, be treated as an empirical relation between ionization potential and spectral class. This mode of regarding the question has the advantage of being quite independent of the adopted temperature scale. We merely assume that the sequence of spectral classes is a temperature sequence. The ionization

[5] Chapter IX, p. 133.

158   EVALUATION OF PHYSICAL CONSTANTS

potentials corresponding to lines of known maximum may then be deduced by interpolation.

TABLE XXV

Element	Ionization Potential	Authority
C++	45	Payne
N+	24	Ibid.
N++	45?	Ibid.
O+	32	Ibid.
O++	45	Fowler and Milne
Si	8.5	Menzel, Payne
S+	20	Payne
S++	32	Ibid.
Sc+	12.5	Menzel
Ti+	12.5	Ibid.
Fe	7.5	Ibid.
Fe+	13	Ibid.

The value of T_{max} is dependent on the effective level, and hence upon the excitation potential. Without the introduction of unjustified assumptions, more than one critical potential cannot be deduced from observations of intensity maximum. The excitation potential corresponding to a line could be roughly inferred from the observed maximum, by observing the shift of predicted maximum produced by the level effect (discussed in Chapter IX) if the ionization potential were known. There are, however, no data as yet that could be used in drawing inferences of this kind.

DURATION OF ATOMIC STATES

The successful application of theory to the astrophysical determination of the life of an atom requires the fulfilment of special conditions. The requirements of the idea developed by Milne[6] demand that the atom shall exist in appreciable quantities in only two states simultaneously. This condition is fulfilled by the ionized atoms of the alkaline earth elements, and it is with calcium that the estimates here discussed are concerned.

[6] Milne, M. N. R. A. S., 84, 354, 1924.

The investigation relates to the calcium present in the high-level chromosphere, where, owing to remoteness from the photosphere, thermal ionization is negligible. Photoelectric ionization may be operative in removing the first electron from the calcium atom, but the sun is too deficient in light of wave-length 1040 for second stage photoelectric ionization to be appreciable. The calcium present in the high-level chromosphere is probably largely in the once ionized condition, since an atom once ionized is likely to remain so for a long time, owing to the scarcity of free electrons in the tenuous outer regions of the sun. The present investigation neglects altogether the neutral and doubly ionized calcium atoms, and furthermore assumes that the transfers corresponding to the H and K lines of the $1^2S - 1^2P$ series are the only ones that occur in appreciable quantities. The latter assumption is apparently not accurately fulfilled, as the $1^2P - m^2D$ lines of Ca+ have recently been detected in the high level chromosphere.[7]

In the simple case of the Ca+ atom (neglecting the small number of atoms that are giving rise to the $1^2P - m^2D$ lines) only two states of the atom are possible: the normal state, called by Milne the 1σ state, and the excited, or 1π state. A given atom exists alternately in these two states. If τ be the average time spent in the 1π state, and τ' the average time spent in the 1σ state, the average time spent by an atom in traversing its possible cycle of changes is $\tau + \tau'$. Now τ is connected with the probability of an emission, and τ' with the probability of an absorption. Clearly τ' depends at least partly upon the energy supply, but τ is an atomic constant measuring the readiness with which the atom recovers its normal state after an absorption. It is, in fact, the "average life" evaluated from Milne's equations. The ratio τ/τ', expressing the relative tendencies of Ca+ atoms to emit and to absorb the H and K lines, is the residual intensity at their centers, with respect to the adjacent continuous background.

[7] Curtis and Burns, unpub.

Einstein's theory of radiation [8] is used in evaluating τ/τ' from the relation

$$\frac{\tau}{\tau'} = \frac{\tfrac{1}{2}r}{e^{h\nu/kT}-1}$$

where r is the ratio $\dfrac{\text{line intensity}}{\text{background intensity}}$.

From ordinary quantum principles,

$$(\tau+\tau') = \frac{\tfrac{1}{2}h\nu}{cmg}$$

and both τ and τ' may be derived by eliminating between the two equations.

The only measured quantity in the formula is r, and from the fact that r is the "residual intensity" within an absorption line, we know that it must lie between 0 and 1. Hence a maximum value of 5.4×10^{-8} seconds may be derived for τ. On the insertion of the data given by Schwarzschild [9] for the residual intensity of the H and K lines, 2.6 magnitudes fainter than the continuous background, and corresponding to a value of r equal to 0.11, the deduced value of τ is 0.6×10^{-8} seconds. The agreement of this value with those obtained in the laboratory for the atoms of hydrogen and mercury has been commented upon in a previous chapter.[10]

[8] Phys. Zeit., **18**, **121**, 1914. [9] Sitz. d. Preuss. Ac., **47**, 1198, 1914.
[10] Chapter I, p. 21.

CHAPTER XII

SPECIAL PROBLEMS IN STELLAR ATMOSPHERES

THE greater part of the present work has dealt with the discussion and interpretation of the normal spectral sequence, Bo to Mb, and the main features of the series have been satisfactorily attributed to thermal ionization at high temperatures. Such a discussion must naturally be the first step in the analysis of the stellar atmosphere. When the more general results of observation have been reduced, in some measure, to an orderly system, it becomes possible to consider special problems involving stars or groups of stars, which lie outside the system, or which, though included in the system, display definite abnormalities.

The special problems of stellar spectroscopy are very numerous. We may mention the novae, the Class O stars, B stars that show emission lines, the problem of A-star classification and the peculiar A stars, F stars that display both giant and dwarf spectral characteristics, the classification of the K stars, the apparent splitting into three groups of the spectral sequence for temperatures below 4000°, the problem of the c-stars, the Cepheid variables with their variable spectra, and the variables of long period.

It is not possible, in a work like the present, to touch upon many of these subjects, and the writer has selected for brief discussion three upon which she can contribute new material: the problem of the O stars, the classification of the A stars, and the interpretation of the spectra of the stars that display the c-characteristic.

THE STARS OF CLASS O

The statistically negligible class containing the O stars is placed, at the present stage of investigation, at the top of the stellar sequence. These spectra indicate higher excitation than those of any other class, and ionization theory distributes their temperatures between 25,000° and 40,000°. Their spectra are among the most puzzling encountered in the whole stellar sequence, and theory has hitherto been unsuccessful in suggesting the conditions that produce them.

A hundred and forty non-Magellanic O stars [1] are enumerated in the Draper catalogue, and in addition a small number of apparently faint B stars should probably be transferred to Class O, as Victoria has already done [2] for a group of stars in Monoceros. The O stars have a very definite distribution; they lie either very near the galactic plane, or in one of the Magellanic clouds, or they constitute the nuclei of planetary nebulae.

The O stars other than the nuclei of planetaries have high intrinsic luminosities, but the material is insufficient for a satisfactory estimate of the absolute magnitudes of the non-Magellanic O stars; various indications point to a value at least as high as -4. For the Magellanic O stars, absolute magnitudes as great as -6.7 have been derived.[3] The measured parallaxes of the planetary nebulae, however, give for the nuclei absolute magnitudes [4] in the neighborhood of $+8$. The wide difference in absolute magnitude can merely be pointed out; it has never received adequate explanation.

The masses are presumably very high for the O stars, though but few have been accurately measured. The star B.D. $6°1309$, a spectroscopic binary reported by J. S. Plaskett,[5] has a minimum mass eighty times that of the sun, and the stars 29 Canis Majoris and ι Orionis also appear to be very massive.[6]

The spectra of the stars of Class O differ widely among themselves, but they are signalized by the lines of ionized helium, which are normally observed only in this class and in the nebulae. In addition, the atoms of H, He, $Mg+$, $C++$, $O++$,

[1] E. B. Wilson and Luyten, Proc. N. Ac. Sci., 11, 133, 1925.
[2] J. S. Plaskett, Pub. Dom. Ap. Obs., 2, 287, 1924.
[3] Shapley and H. H. Wilson, H. C. 271, 1925.
[4] Van Maanen, Proc. N. Ac. Sci., 4, 394, 1918.
[5] J. S. Plaskett, Pub. Dom. Ap. Obs., 2, 147, 183, 269, 1922.
[6] J. S. Plaskett, Pub. Dom. Ap. Obs., 2, 287, 1924.

N++, and Si+++ are represented. The atmospheres of these stars are thus in a state of high ionization, which is attributed to high temperature, in harmony with the work already outlined in previous chapters. The spectra of the stars of Class O have been described by W. W. Campbell,[7] Miss Cannon,[8] Wright,[9] H. H. Plaskett,[10] J. S. Plaskett,[11] and the writer,[12] and the material upon which the present discussion is based will be found in the papers quoted.

Many of the O stars, such as τ Canis Majoris, H.D. 150135, H.D.159176, H.D.199579, H.D. 164794, H.D. 167771, H.D.165052, give a pure absorption spectrum, containing the Balmer series of hydrogen, the lines of Si+++, the Pickering and " 4686 " lines of He+, and the N++ line at 4097. The stars are mentioned in order of increasing ionization, with He+ rising in intensity. The other lines mentioned are clearly beyond their maximum, and fall progressively in strength. The stars mentioned probably represent successive steps in a sequence with rising temperature, connecting directly with Class Bo, and ranging from 25,000° to 35,000°.

This sequence of O stars would form a simple and explicable series if it were an isolated group. There are, however, other stars, with spectra so similar to those of the series just quoted that there can be no doubt of a close relationship — they display absorption lines due to the same elements with about the same relative intensities — but emission lines tend to occur in various parts of the spectrum. 29 Canis Majoris, θ Muscae, and ζ Puppis have absorption spectra which resemble those in the sequence just quoted, but at 4650 and 4686 there are emission lines or " bands." The bright lines are so wide and diffuse, in θ Muscae, as to be blended together at their edges, while they are sharp and clear in the other two stars. Between the two

[7] Ast. and Ap., **13**, 448, 1894.
[8] H. A. **28**, 1900.
[9] Lick Pub., **13**, 248, 1918.
[10] H. H. Plaskett, Pub. Dom. Ap. Obs., **1**, 325, 1922.
[11] J. S. Plaskett, Pub. Dom. Ap. Obs., **2**, 287, 1924.
[12] Payne, H. C. 263, 1924.

series just quoted — pure absorption stars and absorption stars with some bright lines, comes δ Circini, a star which has all the characteristics of the first group, and also shows faint emission on the red edge of the absorption lines at 4650 and 4686.

There are other stars, such as λ Cephei, ξ Persei, S Monocerotis, H.D. 152408, H.D.112244, and ι Orionis, that have absorption spectra such as were described for the first group, but which display faint emission lines at the red edge of the hydrogen and helium lines. It is obvious that all the stars so far enumerated may legitimately be classed together, but that there is a very universal tendency for emission lines and bands to appear in them. This tendency is so marked in the stars that are still to be mentioned as to constitute their most salient feature.

In the subgroup of the O stars which are collectively designated the Wolf-Rayet stars, the emission lines are the most conspicuous characteristic of the spectrum. The best known and brightest star of this group is γ Velorum, which possesses an extremely complex spectrum, made up of an absorption spectrum similar to that of δ Circini, and a large number of wide "emission bands." An analysis of the spectrum of γ Velorum has been published by the writer;[13] all the stronger lines of H, He, He+, C++, O++, N++, Si+++, and Mg+ are represented in the spectrum, and a comparatively small number of lines remains unidentified.

Other O stars that have spectra in which the emission lines are the prominent feature are H.D.151932, H.D. 92740, H.D. 93131, H.D. 152270, H.D. 156385, and H.D. 97152. All of these stars, excepting the last, have also absorption spectra displaying the lines of H and He+. The lines of N++ and Si+++ are absent, and these stars are therefore probably at the extreme high-temperature end of the sequence.

The question of absorption in the Wolf-Rayet spectrum is a difficult one, because the bright lines show up before any other feature of the spectrum, while an appreciable continuous background is necessary before absorption can be detected. The

[13] Payne, H. C. 263, 1924.

detection of absorption lines in many stars, such as H.D. 152270, where no absorption had previously been recorded, has resulted from a general survey of spectra that had received exposures sufficient to bring out the continuous background. The writer has been led to the opinion that absorption is a common, if not universal, feature of all the Wolf-Rayet stars, except those classed at Harvard as Ob. This subclass has bright bands that do not coincide with those of the other O stars, and among them absorption lines appear to be exceptional.

It is perhaps to be expected that absorption should normally occur among the Wolf-Rayet stars, as it does among the other classes. In all other stars, the bright lines that appear are the abnormal feature, and are superposed on a normal continuous spectrum crossed by absorption lines. Spectra consisting of bright lines *only* do not occur elsewhere, excepting for the gaseous nebulae. The gaseous nebulae have, presumably, no photosphere, and the continuous background that they sometimes display is probably the result of reflected and transformed starlight; absorption lines appear normally to accompany the existence of a photosphere.

It is clear that the O stars are a very complex group. Those that have pure absorption spectra can be arranged in a series immediately preceding Bo; and those that show a similar absorption spectrum, with faint superposed emission, presumably also fit into the sequence. When a B star shows emission lines (as γ Cassiopeiae does) it is placed in the B class appropriate to its absorption spectrum, with the additional designation " e " to indicate the abnormality, and the same procedure appears to be equally satisfactory for the O stars.

As emission predominates more and more, the spectrum resembles those of the normal members of the sequence less and less. If a star has an absorption spectrum it can always be assigned a place in the sequence, and this method of arrangement appears to be logical. But it is clear that the sequence so formed is no longer physically homogeneous. The stars that have no absorption lines, although some of them have obvious affinities with stars that have absorption spectra, have moreover no place in a sequence formed on the basis of absorption intensities.

It is, of course, possible to devise a self-consistent scheme for the arrangement of a limited number of the O stars, and such a scheme is, for many purposes, both desirable and convenient. It is, however, exceedingly hard to know where the division should be drawn between " absorption " and " emission " stars. Perhaps the most satisfactory plan is to treat all O stars as a *sequence*, with special comment for the large number of them that require it.

The Class A Stars

(a) *The Balmer Lines.* — The spectra of the A stars are dominated by the Balmer series of hydrogen, which, with the exception of the H and K lines in the cooler stars, are stronger at A0 than any other line seen in the stellar sequence. The maximum of the Balmer series has been stated [14] to occur at A0, and this value was used by Fowler and Milne [15] in calibrating their temperature scale based upon ionization theory. It is in accordance with theory that the subordinate lines of hydrogen, with ionization potential 13.54 volts, and excitation potential 10.15 volts, should have their maximum at about 10,000°.

The position of the maximum can be placed elsewhere by the use of special stars in estimating the line-intensities. The intensity of the hydrogen lines is in fact unusually difficult to determine, as they differ from star to star in width, wings, and probably also in central intensity. Using a series of individual stars, Menzel [16] placed the maximum at A3, with the note that " on the average the lines seem to be strongest in Classes A2 to A4, but the mean intensity is often greatly exceeded in certain A5, A0 and even B8 stars." A general study of the Class A spectra confirms the statement that the mean intensity at maximum is often exceeded for individual stars in other classes, and the writer is inclined to be of the opinion that no significant

[14] Preface, Henry Draper Catalogue.
[15] M. N. R. A. S., **83**, 403, 1923.   [16] H. C. 258, 1924.

THE STARS OF CLASS A   167

maximum can be derived from a limited number of estimates. The maximum given in the Henry Draper Catalogue is the product of the examination of an enormous number of very short dispersion plates, and is entitled to a greater weight than any other. In the estimation of such strong lines, the width and especially the wings are likely to affect the estimates extensively, and short dispersion plates probably reduce the difficulty, and permit of the greatest possible accuracy. It must be emphasized that the maximum given in the Henry Draper Catalogue cannot be superseded by measures made on an arbitrary selection of stars, such as is used when stars bright enough to be photographed with (say) two objective prisms are discussed, for it is a generalization from the most complete data hitherto examined, or to be examined for some time to come. The nonhomogeneity of the A classes, presently to be discussed, includes wide variations in the widths of the hydrogen lines, and renders unnecessary any attempt to correct the hydrogen maximum at A0, which appears to be of a statistical nature.

(b) *Metallic Lines and Band Absorption.* — The metallic lines, which become so conspicuous in intensity in the later classes, appear in the types immediately succeeding A0, and increase progressively in strength as cooler classes are approached. In general, all the related lines belonging to any one element appear at the same class, although sometimes the fainter components of metallic multiplets are not seen until the stronger components have attained a considerable intensity. For example the weaker lines of an element that is seen at A0 may not appear till A2. The disparity in intensity between the components of a multiplet is usually not so great at appearance as at maximum. The relative strengths of unblended lines conform at maximum with the laboratory intensities, to an extent that raises questions as to the degree of saturation [17] of the more intense components. It seems that none of the metallic lines, excepting those of calcium, are greatly oversaturated, even at maximum, to judge from the relative intensities of related lines at that point.

[17] Chapter IV, p. 52.

It is within the A stars that the first signs of band absorption appear. The G band is seen in some A stars, and a drop in the continuous spectrum of Vega around 4160 has been ascribed [18] to the cyanogen band headed at 4215. Similar "band" absorption can be traced in other stars of Class A, and is even seen as early as B0. Identification of the cyanogen band headed at 3883, which always accompanies the 4125 band, would confirm the attribution to cyanogen, but the violet band does not seem to have been observed. The wings of H ζ, which are often wide and conspicuous, render it difficult to trace anything of the nature of band absorption near 3885 for an A star.

(c) *The Classification of A Stars.* — Several lines of evidence have indicated that the classes into which the A stars have been divided are not physically homogeneous, and the problem of their classification is one of the future tasks of astrophysics. It is hoped that the writer can in the future make a more complete discussion of the question than is here desirable, and therefore the present treatment is to be considered merely suggestive and tentative. The material quoted is slight, and must be increased before conclusions can be justified.

It has been suggested [19] that a one-dimensional arrangement will not suffice for the classification of the A stars. The spectra have been classified, at least for the hotter A types, by the strength of the H and K lines of Ca+. With the dispersion used in making the Henry Draper Catalogue, these lines constitute the conspicuous difference between one spectrum and another, and are the obvious criterion of class. If the spectra are classed by the strength of these lines alone, the classification is of course quite unambiguous, and for a one-dimensional sequence of spectra it would have been ideal. That the classes so formed are *not* homogeneous [20] indicates that some second variable must be described in a satisfactory classification, and that the strength of no one line could have been used with any greater success than

[18] Shapley, H. B. 805, 1924.
[19] Shapley, Rep. Spectr. Class. Com., I. A. U., 1925.
[20] See Appendix.

that of the H and K doublet. Further, practical difficulty in duplicating the classifications has been caused by the fact that the H and K lines are so far into the violet that they do not appear at all on many slit spectra, such as those used at Mount Wilson, and when other criteria are chosen for classification, it is likely that the results will deviate somewhat from those of the Henry Draper Catalogue.

By analogy with what is observed in other types, it has been suggested that the range in line-sharpness that is found within a given class among the A stars is an effect of absolute magnitude, and the sharpness of the hydrogen lines has indeed been used at Mount Wilson [21] as a quantitative measure of luminosity. From an analysis of the widths of hydrogen lines made by Miss Fairfield,[22] it appears that the line sharpness may be used to distinguish A stars of the highest luminosity from those of the lowest, but that it cannot be used for the accurate estimation of absolute magnitude between those limits.

The special problem of classifying the A stars is only in its initial stage. That the present system is inadequate is certain, but as yet no satisfactory alternative has been proposed. The direction in which work should be pursued is, in this instance, probably the study of the differences between individual spectra. As the problem appears to hinge on the presence of abnormalities within a given class, it is of especial importance to examine the frequency, magnitude, and nature of these abnormalities.

(d) *Silicon and Strontium Stars.* — There are among the A stars two small groups of especial interest — the so-called " silicon " and " strontium " stars. These occur chiefly in A0, A2, A3, and are distinguished by the unusual intensity of the lines 4128 and 4131 of ionized silicon, and the lines 4077 and 4215 of ionized strontium. Such stars are regarded in the Henry Draper Catalogue as definitely abnormal, and are individually mentioned in the Remarks. The strontium stars in classes later

[21] Mt. W. Contr. 244, 1922; 262, 1923.
[22] H. C. 264, 1924; cf. Lindblad, Ap. J., **59**, 305, 1924.

than Fo are apparently ordinary high-luminosity stars, and the line-intensity is involved in the well known absolute magnitude effect.

The absolute magnitudes of the strontium stars have been supposed, on general grounds, to be very high, but an examination of the proper motions indicates that this is perhaps not even generally true. The well known F star α Circini is apparently a dwarf,[23] and γ Equulei has similar spectral peculiarities and proper motion. Examples might be multiplied, but there is not enough material at present available for a full discussion, and from what has already been said it is evident that the strontium stars constitute no ordinary absolute magnitude problem, although the condition that produces strong strontium lines in some dwarf stars may be something, like low surface gravity, that also prevails in stars of high luminosity.

There are too few parallaxes, proper motions and radial velocities for significant statistical treatment of the silicon stars, and still less material for the strontium stars; but the galactic distributions of both classes indicate that their absolute magnitudes are at least not extremely high. There does not at present appear to be sufficient justification for the statement that these stars are "distinctly brighter than the average."[24] Their brightness would rather seem to be about the same as that of a normal A star.[25]

The silicon and strontium stars raise spectroscopic difficulties that differ somewhat in the two cases. Most of the silicon stars occur at or near Ao, where the Si+ lines are normally at maximum intensity. On the other hand, Sr+ has its maximum at Class K2 or K5, but the intensity in such A stars as ω Ophiuchi and θ_1 Microscopii is as great as it is in these types. The strontium problem illustrates the general conceptions underlying the methods of estimating line-intensities, and will therefore be discussed in slightly more detail.

[23] Shapley, H. B. 798, 1924; Luyten, H. C. 251, 1924.
[24] Rep. Spectr. Class. Com., I A. U., 1922.
[25] Luyten, H. B. 797, 1924.

Abnormal intensity of a spectrum line can be attributed to (1) blending, (2) unusual conditions, or (3) abnormal abundance. These conditions will be discussed in order.

(1) *Blending.* — Blending, excepting where lines are spectroscopically resolved, can only be detected indirectly, by examining the behavior of other lines belonging to the same spectral series as the line in question. If the relative intensities of all the lines in the series are the same as those found in the laboratory, and if changes of intensity from class to class affect all the lines of a series equally, it may be inferred that blending is not a serious disturbing factor and that the abnormal intensity is due to other causes. The close correlation between the stellar intensities of 4215 and 4077, the components of the principal doublet of ionized strontium, in the different spectral classes, leaves little doubt that these lines are effectively unblended in the A stars, although the difference of spectral class for the maximum of the two lines (K2, Ma) and the presence of a solar iron line at 4215, suggest a blend for the latter in stars cooler than about F5. It is also to be remarked that the head of a "cyanogen" band falls at 4216.

(2) *Abnormal Conditions.* — Abnormal conditions permit of no direct observational test, but it would be anticipated that they would also affect other lines to a degree greater than is observed. The change of temperature that would be required to raise the Sr+ lines to their maximum strength at A0 (10,000°) would be a fall of about 5000°, which is quite inadmissible, for the resulting change of spectrum would produce a K star. The required change of pressure is also too great to be possible: this subject cannot profitably be discussed here, and reference should be made to Chapter X. The existence of a strontium cloud has been suggested [26] by analogy with the "calcium cloud," and might possibly provide an explanation, as it would furnish a low temperature for the strontium without unduly lowering the temperature for the star in general. The observation of stationary strontium lines would materially strengthen

[26] J. S. Plaskett, Pub. Dom. Ap. Obs., 2, 335, 1924.

this argument, but they have not so far been recorded. The fact that the strontium stars are scattered, and not concentrated in any one part of the sky, reduces the probability of this suggestion.

(3) *Abnormal Abundance.* — Abnormal abundance has been progressively abandoned as an explanation of the various phenomena of stellar spectra, and that it is the true interpretation of strontium peculiarities seems somewhat unlikely. For the silicon stars, unusual abundance is probably an untenable hypothesis, since the great strength of the Si+ lines is apparently not accompanied by increase in the silicon line, which should presumably occur if pure abundance is the cause of the increased strength of the ionized silicon lines. Abnormal strength of silicon in the cooler stars, doubly ionized silicon in the early B stars, or triply ionized silicon in the O stars, has not been observed, and it is not very probable that, if silicon is unevenly distributed in the universe, the irregularity would be revealed in stars at one temperature only.

Such considerations point to the problem of the silicon and strontium stars as one involving the atom and its energy supply, rather than an abnormal distribution of the element in question. It is likely that the problem of classifying the A stars will be elucidated by a more detailed study of the silicon and strontium lines. The behavior of strontium appears, however, in some cases, to warrant the description of "abnormal," and it may be that the first step in the A star problem will be the elimination from the general classification of spectra such as those of the strontium stars. The present writer inclines to the belief that the silicon and strontium stars will be included in the normal A star classification, when such a one is satisfactorily devised.

(e) *Peculiar Class A Stars.* — Among the A stars there are three which appear to be of special interest.

The star α Andromedae, designated Aop in the Draper Catalogue, has been shown to display enhanced lines of manganese, broadly winged, and of unusual strength.[27]

[27] Lockyer and Baxandall, Proc. Roy. Soc., **77A**, 550, 1906.

The star α Canum Venaticorum has been the subject of extensive work.[28, 29] The chief point of interest concerning it is the occurrence of lines ascribed to the rare earths.[30] The spectra of these elements are so rich in lines that spurious coincidences are certain to occur, but comparisons with the spectrum of α Cygni and of the chromosphere suggest that the strongest lines of europium and terbium are indeed represented. From general ionization principles it would appear that enhanced spectra are probably involved, but until series relations are known it is not possible to discuss the subject further.

The super-giant, or c-star, α Cygni, Class A2p, has probably greater possibilities for the stellar spectroscopist than any other star, as its spectrum is peculiarly rich in fine sharp lines, many of which are unidentified. α Cygni is representative of a large class of stars, but it is the only one of them that has an apparent magnitude bright enough to render it readily accessible. The spectrum has been tabulated by Lockyer[31] and by Wright.[32] At the temperatures concerned, the doubly enhanced lines of the metals are to be anticipated, and it is probable that many of the faint unidentified lines in the spectrum of this star are those of twice ionized metallic atoms. The strongest doubly enhanced lines of the metals fall, as is well known, in the ultra-violet. α Cygni contains the lines of the $1^2P - 3^2D$ series of neutral helium, the most persistent lines of the element, and this is significant in view of the extremely low pressure that is assigned to the atmosphere of the star on the basis of absolute magnitude.

THE C-STARS

The c-characteristic was first used by Miss Maury[33] to designate stars, found in several spectral classes, that have marked spectral peculiarities. The intensities of some of the metallic

[28] Belopolsky, Pub. Ac. Imp. St. Pet., 6, 12, 1913; Pulk. Bul., 6, 10, 1915.
[29] Lockyer and Baxandall, Proc. Roy. Soc., 77A, 550, 1906.
[30] Kiess, Pub. Obs. Mich., 3, 106, 1923.
[31] Lockyer, Pub. Sol. Phys. Com., 1904.
[32] Wright, L. O. B. 332, 1921.
[33] A. C. Maury, H. A., 28, 79, 1897.

SPECIAL PROBLEMS

lines, chiefly those of ionized atoms, are greatly strengthened for the class, and other lines, chiefly those of the neutral atom, are weakened. The G band becomes markedly discontinuous,

TABLE XXVI

Wave Length	Wave Length	Atom	Series	Wave Length	Wave Length	Atom	Series
3758.8	3757.68	Ti+	$1^2F - 1^2F'$	4325.2	4314.98	Ti+	$1^4P - 1^4D'$
	3759.30	Ti+	$2^2D - 2^2D'$	4321.1		Ti+	
3856.2				4337.6	4337.92	Ti+	$1^2D - 1^2D'$
3863.2				4374.7		Ti+?	
3900.7	3900.53	Ti+	$1^2G - 1^2G'$	4395.3	4395.04	Ti+	$1^2D - 1^2F'$
3913.6	3913.45	Ti+	$1^2G - 1^2G'$	4400.2	4399.77	Ti+	$1^2P - 1^4D'$
4003.0	4002.09	Fe+	$2^4P - 1^4P'$	4417.9	4417.71	Ti+	$1^4P - 1^2D'$
4009.4				4444.0	4443.80	Ti+	$1^2D - 1^2F'$
4012.6	4012.40	Ti+	$1^2F - 1^4G'$	4450.6	4450.49	Ti+	$1^2D - 1^2F'$
4024.8	4025.13	Ti+	$1^2F - 1^4G'$	4469.5	4468.14	Ti+	$1^2G - 1^2F'$
4028.5	4028.35	Ti+	$2^2G - 2^2F'$	4471.8	4472.93	Fe+	$2^4F - 1^4F'$
4030.8				4409.6	4489.21	Fe+	$2^4F - 1^4F'$
4053.8	4053.84	Ti+	$2^2G - 2^2F'$	4491.6	4491.41	Fe+	$2^4F - 1^4F'$
4077.9	4077.71	Sr+	$1^2S - 1^2P$	4501.5	4501.27	Ti+	$1^2G - 1^2F'$
4122.8	4122.64	Fe+	$2^4P - 1^4F'$	4508.5	4508.29	Fe+	$2^4F - 1^4D'$
4128.1	4128.	Si+		4515.4	4515.34	Fe+	$2^4F - 1^4F'$
4131.4	4131.	Si+		4520.3	4520.24	Fe+	$2^4F - 1^4F'$
4143.9				4522.9	4522.64	Fe+	$2^4F - 1^4D'$
4173.6	4173.47	Fe+	$2^4P - 1^4D'$	4534.2	4533.97	Ti+	$1^2G - 1^2D'$
4179.5	4178.87	Fe+	$2^4P - 1^4F'$		4534.17	Fe+	$2^4F - 1^4F'$
4187.6				4556.0	4555.90	Fe+	$2^4F - 1^4D'$
4215.7	4215.52	Sr+	$1^2S - 1^2P$	4558.9			
4233.6	4233.16	Fe+	$2^4P - 1^4D'$	4564.0	4563.77	Ti+	$1^2P - 1^2D'$
4271.7		Fe+?		4584.0	4583.84	Fe+	$2^4F - 1^4D'$
4288.1	4287.88	Ti+	$1^2D - 1^2D'$	4586.			
4294.3	4294.10	Ti+	$1^2D - 1^2D'$	4619.2	4620.52	Fe+	$2^4F - 1^4D'$
4297.1	4296.56	Fe+	$2^4P - 1^4F'$	4629.9	4629.33	Fe+	$2^4F - 1^4F'$
4314.4	4312.88?	Ti+	$1^4P - 1^4D'$	4657.0			

and heavy blends at 4072, 4077, become conspicuous. The spectrum of a c-star is unmistakable in appearance.

The foregoing tabulation contains a list of the lines that are strongly enhanced in the spectra of the c-stars. Successive columns give the approximate wave-length, taken from Miss

Maury's original list, the laboratory wave-length of the line with which the stellar line is identified, the atom, and the series relations.

The preponderating characteristic lines are clearly those of ionized iron and ionized titanium.[34] All the strong lines of these atoms are found in the c-star spectrum, and they are there stronger than in any other class. It is found that spectra possessing the c-character have in general unusually sharp and narrow lines. It is probable that the lines in the spectrum of a

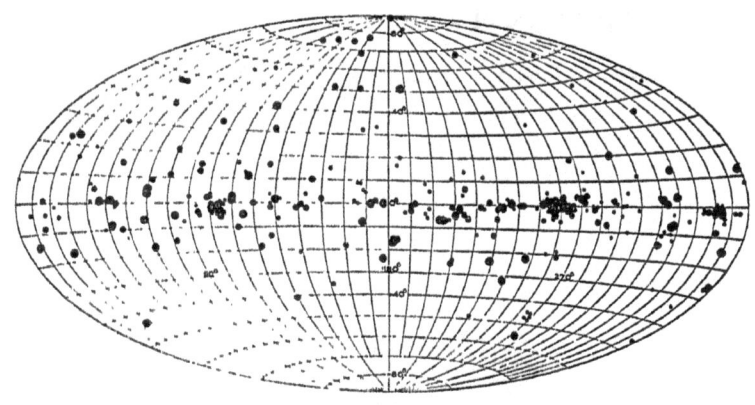

FIGURE 9

Galactic distribution of stars mentioned in the Draper Catalogue as having narrow lines. Four sizes of dots indicate stars of different apparent magnitudes; brighter than 5.0; 5.0 − 6.0; 6.0 − 8.0; and fainter than 8.0, respectively.

c-star are actually stronger, as well as sharper, than the corresponding lines in a star of the same class and lower luminosity.[35] This phenomenon is connected with the question of the effective optical depth of the photosphere, and is discussed in Chapter IX.

It was first pointed out by Hertzsprung [36] that the c-character marks out a class of stars with distinct physical properties — extremely small parallaxes and proper motions, strong galactic concentration, and, accordingly, very high luminosity and vol-

[34] Russell, Ap. J., in press.
[35] Stewart, Phys. Rev., 22, 324, 1923; Russell and Stewart, Ap. J., 59, 197, 1924.
[36] A. N., 179, 374, 1908; A. N., 192, 262, 1912.

ume, and low density. The last feature furnishes an interpretation of the spectral peculiarities (see Chapter X).

The reality of the c-character has been questioned owing to a misapprehension as to its criteria.[37] Fine lines always accompany the c-character, but they may be present without it. The star h Ursae Majoris is a case in point. It is listed in the Henry Draper Catalogue as having narrow lines, a remark that usually indicates the presence of the c-character. Actually the star appears to be a dwarf, of Class F0, with considerable proper motion. Although the lines are narrow and sharp, the spectrum has not the very typical appearance of a c-star.

[37] Harper and Young, J. R. A. S. Can., 18, 9, 1924.

CHAPTER XIII

THE RELATIVE ABUNDANCE OF THE ELEMENTS

THE relative frequency of atomic species has for some time been of recognized significance. Numerous deductions have been based upon the observed terrestrial distribution of the elements; for example, attention has been drawn to the preponderance of the lighter elements (comprising those of atomic number less than thirty), to the "law of even numbers," which states that elements of even atomic number are far more frequent than elements of odd atomic number, and to the high frequency of atoms with an atomic weight that is a multiple of four.

The existence of these general relations for the atoms that occur in the crust of the earth is in itself a fact of the highest interest, but the considerations contained in the present chapter indicate that such relations also hold for the atoms that constitute the stellar atmospheres and therefore have an even deeper significance than was at first supposed. Data on the subject of the relative frequency of the different species of atoms contain a possible key to the problem of the evolution and stability of the elements. Though the time does not as yet seem ripe for an interpretation of the facts, the collection of data on a comprehensive scale will prepare the way for theory, and will help to place it, when it comes, on a sound observational basis.

The intensity of the absorption lines associated with an element immediately suggests itself as a possible source of information on relative abundance. But the same species of atom gives rise simultaneously to lines of different intensities belonging to the same series, and also to different series, which change in intensity relative to one another according to the temperature of the star. The intensity of the absorption line is, of course, a very complex function of the temperature, the pressure, and the atomic constants — a matter that has been discussed in detail in the preceding seven chapters.

The observed intensity can therefore be used *directly* for only a crude estimate of abundance. Roughly speaking, the lines of the lighter elements predominate in the spectra of stellar atmospheres, and probably the corresponding atoms constitute the greater part of the atmosphere of the star, as they do of the earth's crust. Beyond a general inference such as this, few direct conclusions can be drawn from line-intensities. Russell [1] made the solar spectrum the basis of a discussion in which he pointed out the apparent similarity in composition between the crust of the earth, the atmosphere of the star, and the meteorites of the stony variety. The method used by him should be expected, in the light of subsequent work, to yield only qualitative results, since it took no account of the relative probabilities of the atomic states corresponding to different lines in the spectrum.

UNIFORMITY OF COMPOSITION OF THE STELLAR ATMOSPHERE

The possibility of arranging the majority of stellar spectra in homogeneous classes that constitute a continuous series, is an indication that the composition of the stars is remarkably uniform — at least in regard to the portion that can be examined spectroscopically. The fact that so many stars have *identical* spectra is in itself a fact suggesting uniformity of composition; and the success of the theory of thermal ionization in predicting the spectral changes that occur from class to class is a further indication in the same direction.

If departures from uniform distribution did occur from one class to another, they might conceivably be masked by the thermal changes of intensity. But it is exceedingly improbable that a lack of uniformity in distribution would *in every case* be thus concealed. It is also unlikely, though possible, that a departure from uniformity would affect equally and solely the stars of one spectral class. Any such departure, if found, would indicate that the presence of abnormal quantities of certain elements was

[1] Russell, Science, 39, 791, 1914.

an effect of temperature. This explanation appears, however, to be neither justified nor necessary; there is no reason to assume a sensible departure from uniform composition for members of the normal stellar sequence.

Marginal Appearance of Spectrum Lines

Fowler and Milne [2] pointed out that the "marginal appearance," when the line is at the limit of visibility, is a function of the abundance of the corresponding atom. For this reason their own theory, which dealt not with the marginal appearance but with the maximum of an absorption line, was capable of a more satisfactory observational test than Saha's. It is possible, as shown below, to extend the Fowler-Milne considerations and to use the observed marginal appearances as a measure of relative abundance.

The conditions for marginal appearance must first be formulated. When a strong absorption line is at maximum, the light received from its center comes from the deepest layer that is possible for the corresponding frequency. The actual depth depends, as was pointed out in Chapter IX, upon the number of absorbing atoms per unit volume, and upon the atomic absorption coefficient for the frequency in question. The suggestions that were put forward in the chapter just quoted indicate that different lines, at their maxima, arise from different "effective levels," the more abundant atoms appearing, other things being equal, at higher levels.

As an absorption line is traced through the classes adjacent to the one at which it attains maximum, it begins to diminish in intensity, owing to the decrease in the number of suitable atoms. If the line is very intense, the first effect of the fall in the number of suitable atoms is a reduction in the width and wings. As the number of suitable atoms per unit volume decreases further, a greater and greater thickness of atmosphere is required to produce the same amount of absorption, and accordingly the line originates deeper and deeper in the atmosphere of the star. As

[2] R. H. Fowler and Milne, M. N. R. A. S., 83, 403, 1923.

the "effective level" falls, the temperature of the layer that gives rise to the line increases, owing to the temperature gradient in the stellar reversing layer. The observed fall in the intensity of the line is caused both by the reduction in the number of suitable atoms, and by the decreased contrast between the line and the background. The former cause predominates for strong (saturated) lines, and the latter for weak (unsaturated) lines.

As the atoms suitable to the absorption of the line considered decrease in number, the effective level from which the line takes its origin falls, and ultimately coincides with the photosphere (the level at which the *general* absorption becomes great enough to mask the *selective* absorption due to individual atoms). The line then disappears owing to lack of contrast. Immediately before the line merges into the photosphere (the approximate point estimated as "marginal appearance"), *all* the suitable atoms above the photosphere are clearly contributing to the absorption; in other words the *line* is unsaturated. The position in the spectral sequence of the marginal appearance of a line must then depend directly upon the *number of suitable atoms above the photosphere*; considerations of effective level are eliminated. Hence a constant P_e is used on page 184.

The conditions at maximum and marginal appearance of a line in the spectral sequence are to some extent reproduced for an individual absorption line at the center of the line and at the edge of its wing. A hydrogen line displays wings that may extend to thirty Angstrom units on either side of the center. The energy contributing to the wings is evidently light coming from hydrogen atoms with a frequency that deviates somewhat from the normal. Atoms with small deviations are more numerous than atoms with large deviations, and therefore the light received from them originates in a higher effective level. The line center corresponds to the highest level of all. At points far out upon the wings, lower and lower levels are represented, until, where the line merges into the continuous background, the level from which it originates coincides with the photosphere, and the "marginal appearance" of the line (if it may so be called) is reached. Accurate photometry of the centers and wings of strong absorption lines would seem to have an important bearing on the structure of the stellar atmosphere, as it would provide an immediate measure of the factor that produces the deviations from normal frequency. The success of parallel work in the laboratory [3] indicates that intensity distribution should be amenable to observation and to theory.

Observed Marginal Appearances

The spectral class at which a line is first or last seen is obviously, to some extent, a function of the spectrosopic dispersion used, for, with extremely small dispersion, many of the fainter lines fail to appear at all. A line will also probably appear somewhat later, and disappear somewhat earlier, with small than with large dispersion. It is therefore a matter of some difficulty to obtain measures of marginal appearance that shall be absolute, but the present discussion neither assumes nor requires them. The method used is designed for the estimation of *relative* abundances, and all that is required of the data is that they shall be mutually consistent.

In order to attain the maximum degree of consistency, the estimates used in this chapter were derived chiefly from the two series of plates mentioned in Chapter VIII. All the plates used were made with the same dispersion (two 15° objective prisms) and were of comparable density, and of good definition. The data furnished by the writer's own measures were supplemented by some estimates derived by Menzel[4] from a similar series of plates, of the same dispersion and comparable quality. The estimate of the marginal appearance of potassium was very kindly suggested by Russell from solar observations.

The observed marginal appearances of all the lines that are available are summarized in the table that follows. Successive columns contain the atomic number and atom, the series relations, the wave-length of the line used, and the Draper classes at

[3] Harrison, unpub.   [4] H. C. 258, 1924.

which the line is observed, respectively, to appear, to reach maximum, and to disappear. Asterisks in the last column denote the ultimate lines of the neutral atom, which are strongest

TABLE XXVII

Atom	Series	Line	Classes	Atom	Series	Line	Classes
1 H	$1S - 2P$	4340	$- A_3 -$	22 Ti	$1F - F$	3999	$*\ *\ A_2$
2 He	$1^2P - 3^3D$	4471	$B_9\ B_3\ O$		$1F - G$	4862	$*\ *\ A_2$
	$1S - 2P$	5015	$B_9\ B_3\ O$			4867	$*\ *\ A_2$
	$1P - 4D$	4388	$B_9\ B_3\ O$			4856	$*\ *\ A_2$
He+	$4F - 9G$	4542	$O\ O\ -$		$1^5F - {}^5F$	4536	$-\ -\ A_5$
3 Li	$1^2S - 1^2P$	6707	$*\ *\ -$			4535	$-\ -\ A_5$
6 C+	$2^2D - 3^2F$	4267	$B_9\ B_3\ O$	23 V	$1^6G - {}^6G$	4333	$*\ *\ Fo$
11 Na	$1^2S - 1^2P$	5889	$*\ *\ Ao\ddagger$			4330	$*\ *\ Fo$
		5896	$*\ *\ Ao\ddagger$	24 Cr	$1^7S - 1^7P$	4290	$*\ *\ A_2$
12 Mg	$1^3P - 1^3D$	5184	$-\ ?\ Ao\ddagger$			4275	$*\ *\ A_2$
		5173	$-\ ?\ Ao\ddagger$			4254	$*\ *\ A_2$
		5167	$-\ ?\ Ao\ddagger$		$1^5S - 1^5P$	4497	$-\ M_1\ A_7$
	$1^3P - 2^3D$	3838	$-\ ?\ Ao$	25 Mn	$1^6S - 1^6P$	4034	$*\ *\ A_2$
		3832	$-\ ?\ Ao$			4033	$*\ *\ A_2$
		3829	$-\ ?\ Ao$			4030	$*\ *\ A_2$
Mg+	$2^2D - 3^2F$	4481	$-\ A_3\ Bo$		$1^6D - {}^6D$	4084	$-\ K_2\ A_3$
13 Al	$1^2P - 1^2S$	3962	$*\ *\ Ao$			4041	$-\ K_2\ A_3$
		3944	$*\ *\ Ao$	26 Fe	$1^3F - {}^3G$	4325	$-\ K_2\ A_2$
14 Si		3905	$-\ Go\ A_2$		$1^3F - {}^3F$	4072	$-\ Ko\ Ao$
Si+		4128	$Fo\ Ao\ O$	30 Zn	$1^3P - 1^3S$	4811	$G_5\ Go\ A_7\dagger$
		4131	$Fo\ Ao\ O$			4722	$G_5\ Go\ A_7\dagger$
19 K	$1^2S - 1^2P$	4044	$*\ *\ F_8$	38 Sr	$1S - 1P$	4607	$*\ *\ Fo$
		4047	$*\ *\ F_8$	Sr+	$1^2S - 1^2D$	4078	$-\ K_2\ Ao$
20 Ca	$1S - 1P$	4227	$*\ *\ B_9$	54 Ba+	$1^2S - 1^2P$	4555	$-\ ?\ A_2$
	$1^3P - 2^3D$	4455	$-\ K_2\ Fo$				
Ca+	$1^2S - 1^2P$	3933	$-\ -Bo$				

at low temperatures, and have no maximum. Estimates by Menzel are indicated by a dagger; those marked by a double dagger were taken from dyed plates made with slightly smaller dispersion.

Method of Estimating Relative Abundances

If the physical conception of marginal appearance above outlined is correct, the *number of atoms* of a given kind above the photosphere will practically determine the class at which the corresponding line is last seen.[5] Now at marginal appearance the number of suitable atoms is only a small fraction of the total amount of the corresponding element that is present in the reversing layer, and this fraction is precisely the "fractional concentration" evaluated by Fowler and Milne. If then it be assumed that the number of atoms required for marginal appearance is the same for all elements, the reciprocals of the computed fractional concentrations at marginal appearance should give directly the relative abundances of the atoms.

A few remarks concerning the underlying assumptions may be appropriate. In applying the theory it is assumed that stellar atmospheres are of uniform composition, and that at marginal appearance all lines are unsaturated. These reasonable assumptions have been discussed above, and they are here explicitly restated. The third assumption, that the same number of atoms is represented at the marginal appearance of a line, whatever the element, is by far the most serious. It implies the equality of the absorbing efficiencies of the individual atoms under the conditions involved. This is assumed in default of a suitable correction, but it is not suggested that the use here made of the assumption would imply its universal validity. Its present application is made under conditions of extremely low pressure (1.31×10^{-4} atmospheres), and over a range of temperature from $7000°$ to $10,000°$. Under such conditions the absorbing efficiency of an atom will depend almost entirely upon its energy supply and upon its inherent tendency to recover after undergoing an electron transfer. The pressures are so low that collisions will have no appreciable effect in disturbing the normal recovery of the atoms. The energy supply will vary with the temperature; but with the range of temperature considered the

[5] Payne, Proc. N. Ac. Sci., 11, 192, 1925.

184 RELATIVE ABUNDANCE OF THE ELEMENTS

variation will probably not be very large. The reorganization time of an atom appears to be an atomic constant, and to be of the same order for all atoms hitherto examined in the laboratory or in stellar atmospheres. As a working assumption, then, the equality of the atomic absorption coefficients is assumed with

TABLE XXVIII

Atomic Number	Atom	Log a_r	Atomic Number	Atom	Log a_r	Atomic Number	Atom	Log a_r
1	H	11	13	Al	5.0	23	V	3.0
2	He	8.3	14	Si	4.8	24	Cr	3.9
	He+	12		Si+	4.9	25	Mn	4.6
3	Li	0.0		Si+++	6.0	26	Fe	4.8
6	C+	4.5	19	K	3.5	30	Zn	4.2
11	Na	5.2	20	Ca	4.8	38	Sr	1.8
12	Mg	5.6		Ca+	5.0		Sr+	1.5
	Mg+	5.5	22	Ti	4.1	54	Ba+	1.1

some confidence in the discussion of observed marginal appearances.

As stated above, the relative abundances of the atoms are given directly by the reciprocals of the respective fractional concentrations at marginal appearance. The values of the relative abundance thus deduced are contained in Table XXVIII. Successive columns give the atomic number, the atom, and the logarithm of the relative abundance, a_r.

COMPARISON OF STELLAR ATMOSPHERE AND EARTH'S CRUST

The preponderance of the lighter elements in stellar atmospheres is a striking aspect of the results, and recalls the similar feature that is conspicuous in analyses of the crust of the earth.[6] A distinct parallelism in the relative frequencies of the atoms of the more abundant elements in both sources has already been suggested by Russell,[7] and discussed by H. H. Plaskett,[8] and the

[6] Clarke and Washington, Proc. N. Ac. Aci., 8, 108, 1922.
[7] Russell, Science, 39, 791, 1914.   [8] Pub. Dom. Ap. Obs., 1, 325, 1922.

STELLAR ATMOSPHERE AND EARTH'S CRUST 185

data contained in Table XXVIII confirm and amplify the similarity.

A close correspondence between the percentage compositions of the stellar atmosphere and the crust of the earth would not, perhaps, be expected, since both sources form a negligible fraction of the body of which they are a part. There is every reason to suppose, on observational and theoretical grounds, that the composition of the earth varies with depth below the surface; and the theory of thermodynamical equilibrium would appear to lead to the result that the heavier atoms should, on the average, gravitate to the center of a star. If, however, the earth originated from the surface layers of the sun,[9] the percentage composition of the whole earth should resemble the composition of the solar (and therefore of a typical stellar) atmosphere. But the mass of the earth alone is considerably in excess of the mass of the reversing layer of the sun.[10] Eddington,[11] quoting von Zeipel,[12] has pointed out that an effect of rotation of a star will be to keep the constituents well mixed, so that the outer portions of the sun or of a star are probably fairly representative of the interior. Considering the possibility of atomic segregation both in the earth and in the star, it appears likely that the earth's crust is representative of the stellar atmosphere.

The most obvious conclusion that can be drawn from Table XXVIII is that all the commoner elements found terrestrially, which could also, for spectroscopic reasons, be looked for in the stellar atmosphere, are actually observed in the stars. The twenty-four elements that are commonest in the crust of the earth,[13] in order of atomic abundance, are oxygen, silicon, hydrogen, aluminum, sodium, calcium, iron, magnesium, potassium, titanium, carbon, chlorine, phosphorus, sulphur, nitrogen, manganese, fluorine, chromium, vanadium, lithium, barium, zirconium, nickel, and strontium.

The most abundant elements found in stellar atmospheres,

[9] Jeffreys, The Earth, 1924. [10] Shapley.
[11] Nature, 115, 419, 1925. [12] M. N. R. A. S., 84, 665, 1924.
[13] Clarke and Washington, Proc. N. Ac. Sci., 8, 108, 1922.

also in order of abundance, are silicon, sodium, magnesium, aluminum, carbon, calcium, iron, zinc, titanium, manganese, chromium, potassium, vanadium, strontium, barium, (hydrogen, and helium). All the atoms for which quantitative estimates have been made are included in this list. Although hydrogen and helium are manifestly very abundant in stellar atmospheres, the actual values derived from the estimates of marginal appearance are regarded as spurious.

The absence from the stellar list of eight terrestrially abundant elements can be fully accounted for. The substances in question are oxygen, chlorine, phosphorus, sulphur, nitrogen, fluorine, zirconium, and nickel, and none of these elements gives lines of known series relations in the region ordinarily photographed.

The $1^5S - m^5P$ "triplets" of neutral oxygen, in the red, should prove accessible in the near future; the point of disappearance of these lines would not be difficult to estimate, and they would furnish a value for the stellar abundance of oxygen. The lines of ionized oxygen, which have not yet been analyzed into series, are conspicuous in the B stars,[14] and the element is probably present in large quantities.

Sulphur and nitrogen both lack suitable lines in the region usually studied; the analyzed spectrum of neutral sulphur is in the green and red,[15] or in the far ultra-violet,[16] and the neutral nitrogen spectrum has not as yet been arranged in series. Both sulphur and nitrogen appear, in hotter stars, in the once and twice ionized conditions,[17] and are probably abundant elements in stellar atmospheres.

For the remaining elements, phosphorus, chlorine, fluorine, zirconium and nickel, series relations are not, as yet, available. No lines of phosphorus or the halogens have been detected in stellar spectra, but these elements have not been satisfactorily analyzed spectroscopically, and their apparent absence from the stars is probably a result of a deficiency in suitable lines. Nickel

---

[14] H. C. 256, 1924.
[15] Fowler, Report on Series in Line Spectra, 170, 1922.
[16] Hopfield, Nature, 112, 437, 1923.   [17] H. C. 256, 1924.

STELLAR ATMOSPHERE AND EARTH'S CRUST 187

and zirconium will probably be analyzed in the near future; they are both well represented in stellar spectra, and nickel especially is probably abundant.

The relative abundances, in the stellar atmosphere and the earth, of the elements that are known to occur in both, display a striking numerical parallelism. Table XXIX gives the data for the sixteen elements most abundant in the stellar atmosphere. Successive columns give the atomic number, the atom,

TABLE XXIX

Atomic Number	Atom	Stellar Abundance	Terrestrial Abundance		Abundance Stony Meteorites
			Crust	Whole Earth	
14	Si	5.7	16.2	9.58	11.2
11	Na	5.7	2.02	0.97	0.6
12	Mg	4.2	0.42	3.38	2.8
13	Al	3.6	4.95	2.66	1.1
6	C	3.6	0.21	....	....
20	Ca	2.9	1.50	1.08	0.56
26	Fe	2.5	1.48	46.37	5.92
30	Zn	0.57	0.0011	....	....
22	Ti	0.43	0.241	0.12	....
25	Mn	0.36	0.035	0.06	....
24	Cr	0.29	0.021	0.05	0.29
19	K	0.11	1.088	0.38	0.10
23	V	0.05	0.0133	....	....
38	Sr	0.002	0.0065	....	....
54	Ba	0.005	0.0098	....	....
3	Li	0.0000	0.0829	....	....

the relative stellar abundance, the relative terrestrial abundance (both for the lithosphere, hydrosphere, and atmosphere, and for the whole earth),[18] and the relative abundance in stony meteorites.[19] The figures in the fifth column are derived from Clarke's estimates of the percentage composition of the earth. The composition of the earth has been variously estimated by different

[18] Clarke, U. S. Geol. Surv. Prof. Pap. 132, 1924.
[19] G. P. Merrill, quoted by Clarke, U. S. Geol. Surv. Bul. 491.

investigators, and the resulting figures depend upon theories that cannot be discussed here. The order given by Clarke is based on the assumption of a nickel-iron core.

The numbers expressing the stellar abundance are percentages, calculated on the assumption that the stellar and terrestrial elements form the same fraction of the total material present. This reduces the two columns of numbers to a form in which they are directly comparable, but no great importance is attached to the absolute percentages in the third column.

The method that has here been used is subject to inaccuracy and uncertainty, especially in the estimates of the exact spectral class at which a line is first or last seen. The most that can be expected is that the results will be trustworthy in order of magnitude. It may be seen that the only element for which the stellar and terrestrial values are not of the same order is zinc. Further, it appears that when the estimates for the percentage composition of the *whole* earth are used in the comparison with the stellar values, the agreement is improved in the case of silicon, magnesium, aluminum, manganese, chromium, and potassium; it is about the same for calcium and titanium, is less close for sodium, and markedly poorer for iron.* In the stellar atmosphere and the meteorite the agreement is good for all the atoms that are common to the two, but several important elements are not recorded in the meteorite.

The outstanding discrepancies between the astrophysical and terrestrial abundances are displayed for hydrogen and helium. The enormous abundance derived for these elements in the stellar atmosphere is almost certainly not real. Probably the result

---

* Professor Russell believes that iron is much more abundant, at least in the sun, than calculated above. He writes: "More than half of all the strong winged solar lines are iron lines, and the strength and evident saturation of even the faint satellites in the iron multiplets is remarkable. . . . There are a great many multiplets of nearly equal strength arising from the low triplet F level in iron. . . . Nothing like this happens for the D lines, or for H and K, although it may hold true for the Mg triplets. I should consequently favor multiplying the percentage for iron by a factor of at least 3 and probably 5 — which would put it where it obviously belongs."

UNIFORMITY OF STELLAR ATMOSPHERES 189

may be considered, for hydrogen, as another aspect of its abnormal behavior, already alluded to;[20] and helium, which has some features of astrophysical behavior in common with hydrogen, possibly deviates for similar reasons. The lines of both atoms appear to be far more persistent, at high and at low temperatures, than those of any other element.

The uniformity of composition of stellar atmospheres appears to be an established fact. The quantitative composition of the atmosphere of a star is derived, in the present chapter, from estimates of the "marginal appearance" of certain spectral lines, and the inferred composition displays a striking parallel with the composition of the earth.

The observations on abundance refer merely to the stellar atmosphere, and it is not possible to arrive in this way at conclusions as to internal composition. But marked differences of internal composition from star to star might be expected to affect the atmospheres to a noticeable extent, and it is therefore somewhat unlikely that such differences do occur.

[20] Chapter V, p. 56.

CHAPTER XIV

THE MEANING OF STELLAR CLASSIFICATION

IT is not necessary to discuss the possibility or desirability of classifying stellar spectra. Both have been adequately demonstrated by Miss Cannon in the Henry Draper Catalogue,[1] which contains the classification that has been accepted as standard.[2] The catalogue will undoubtedly long remain the authoritative source of spectral data for the major part of the stars bright enough to be accessible to the spectroscopist. The uses of the material that it contains are so numerous and so direct that the basis and meaning of the classes seem to deserve attention.

In classifying a number of objects, an attempt should be made to select criteria that will distribute the material into the most natural groups. A classification devised with one point of view will not necessarily appear natural from another, and the best that can generally be done is to select the standpoint that seems to be the most important. From all other standpoints the classification is empirical, and must be treated as such. It seems necessary to emphasize this empiricism with regard to the classifying of stellar spectra, for reference is often made to the Henry Draper Classification as though it had a theoretical, even an evolutionary, basis, whereas it is essentially arbitrary. It is true that a classification based on theoretical principles is very desirable, but at present there is no adequate physical theory on which to found one.

The essential feature of the Draper classification is that it aims at classing together similar spectra, relying on general appearance, and not on the measurement of any one line or group of lines. This has the advantage of distributing the material in the most natural groups possible, and a disadvantage in that

[1] H. A., 91–99.   [2] Rep. I. A. U., Rome, 1922.

different observers may find it difficult to be sure that their criteria are identically weighted.

That the original aim was empirical and not theoretical is clear from the introduction to the first extensive list of spectra classified according to the Draper system:[3] "It was deemed best that the observer should place together all stars having similar spectra and thus form an arbitrary classification rather than be hampered by any preconceived theoretical ideas." The present classification was the natural outcome of such a procedure. As A. Fowler has remarked,[4] "the Draper classification is based essentially on the observed spectral lines, and in reality may be regarded as independent of any other consideration whatsoever. Even if we did not know the origin of a single line in the stellar spectra, it is probable that we should have arrived at precisely the same order."

The descriptions that are contained in the preface to the Henry Draper Catalogue, and which have long been classical, were designed to describe the salient features of the groups that had been formed. It is only in a somewhat restricted sense that they constitute the criteria for those groups. The descriptions were compiled from the spectra of apparently bright stars of the classes involved, but the greater number of the spectra actually classified are taken with such short dispersion that all except the very strongest lines are difficult to distinguish, and are certainly not susceptible of accurate *measurement*. This fact should affect the standpoint of those who criticize the "multiple nature" of the Draper criteria. A portion of one of the plates used in the classification is herewith reproduced with no magnification. This photograph should make it apparent to anyone familiar with the use of spectra that the classification of stars is very largely a *practical* problem.

Instead, then, of examining the possible merits of the best theoretical classification system, it appears to be more useful to examine the physical implication of the most representative classification that it has been found possible to make in practice.

[3] H. A., **28**, 131, 1901. [4] Observatory, **38**, 381, 1915.

The fact that the Draper system is so representative has been regarded as one of its great merits, and has rightly placed it in the authoritative position that it occupies.

When a group of stars is being studied for a special purpose, it is often found that the Draper classes are not fine enough to subdivide the material usefully. In such cases reclassification is often essential. It has sometimes been suggested that this indicates that the Draper classes are inadequate; but it must be recollected that, for the greater part of the material contained in the Catalogue, finer classification would have been impossible, and the subclasses in use today represent the practical survival from a far larger number, which were originally thought to be usable. Actually the stars represent a continuous gradation from class to class, and in classifying it is only possible to use the smallest distinguishable steps, which will obviously be smaller, the larger the dispersion. When it is found necessary to reclassify the stars more finely in a special investigation, as in the Harvard or Mount Wilson work on spectroscopic parallaxes,[5] one or more measurable criteria are selected and used as a basis, but standard stars classified at Harvard are used to define the scale. These measured or closer classifications, while essential for the purpose for which they were designed, have no theoretical advantage over the Draper system (on which they are ultimately founded), and do not, as is sometimes inferred,[6] indicate that the latter is in error.

Although devised with no theoretical basis, the Draper classification has long been recognized as classifying something physical, and the fact that the majority of the stars had been ranged by it in a single sequence suggested that a single variable was principally involved. From general theoretical considerations it could have been predicted that this variable was probably the temperature, but, in addition, the observational evidence that this was the case was immediately convincing. In the words

[5] Mt. W. Contr. 199, 1918.
[6] Harper and Young, J. R. A. S. Can., 18, 9, 1924.

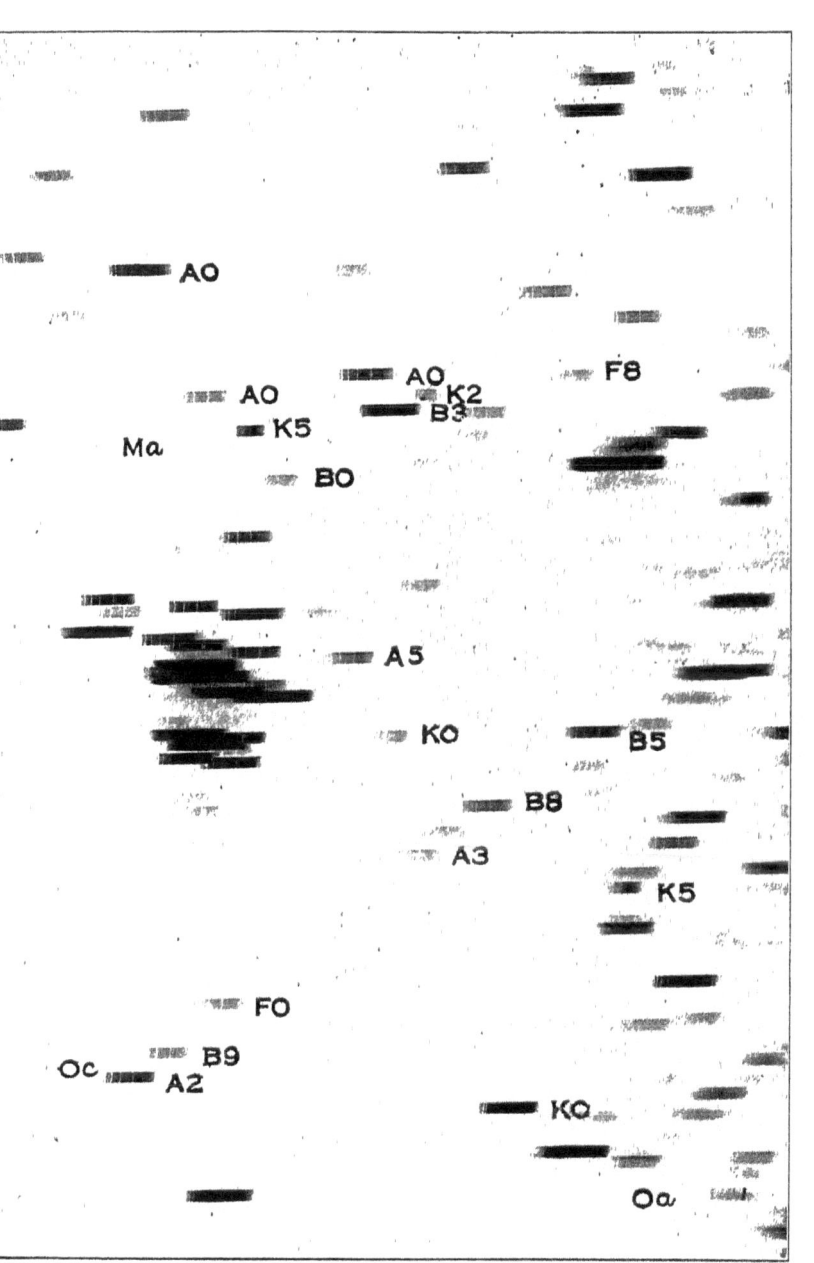

HOMOGENEITY OF CLASSES

of A. Fowler,[7] "... the typical stars not only increase in redness in passing through the sequence, but successive Draper classes correspond to nearly equal increments of redness as measured by the color index."

The preceding eight chapters review the arguments and the observations that have established the connection between the spectrum of a star and its temperature. From an examination of the data there given it becomes clear that what the Draper system classifies is essentially the degree of thermal ionization. A. Fowler, in fact, makes the illuminating distinctions of "arc" (N to G), "spark" (F to A), and "superspark" (B onwards) stars.

The table that follows contains, in concise form, the chief features by which the type stars of each class are to be recognized, although it is again emphasized that these were not actually measured as criteria for the Draper classes. The lines characteristic of each class serve, however, to specify its degree of thermal ionization.

The homogeneity of the spectra in a given class is striking, and the fact that large numbers of stars display exactly similar spectra has a significance — considered in another chapter [8] — to which the classification problem cannot do more than call attention. The similarity of the spectra becomes the more striking when it is remembered that the range of conditions embraced within any one class is very wide; the ratio in mean density may be as great as 10^9 between stars of the same class but of differing absolute magnitude.[10]

The close spectral similarity between giants and dwarfs, in spite of the great differences in physical conditions, should not, however, be misinterpreted. The observed facts are in exact accordance with what might have been anticipated. In the first place, thermal ionization is governed by the surface gravity, and only indirectly by the mean density.[11] It is shown in Chapter III that the range in surface gravity is far smaller than the

---

[7] Observatory, 38, 381, 1915. [8] Chapter XIII, p. 178.
[9] Observatory, 38, 381, 1915. [10] Chapter III, p. 36.
[11] Chapter III, p. 35.

TABLE XXX

Characteristic Lines

Class											
O	H	He	Si+++	C++	N++	He+					
Bo	H	He	Si+++			He+	O+				
Br	H	He	Si+++				O+*				
B2	H	He*									
B3	H	He									
B5	H	He						Si+	Ca+		
B8	H	He						Si+	Ca+	Mg+	
B9	H	He						Si+	Ca+	Mg+	
Ao	H*							Si+*	Ca+	Mg+	
A2	H								Ca+	Mg+*	Ca	Fe
A3	H								Ca+	Mg+	Ca	Fe
A5	H								Ca+		Ca	Fe
Fo	H								Ca+		Ca	Fe
F2	H								Ca+		Ca	Fe	Ti
F5	H								Ca+		Ca	Fe	Ti
Go	H								Ca+		Ca	Fe	Ti	G bd.
Ko	H								Ca+*		Ca	Fe*	Ti*	G bd.
Ma	H								Ca+		Ca*	Fe	Ti	G bd.
Mb	H								Ca+		Ca	Fe	Ti	G bd*	TiO₂
Md	H								Ca+		Ca	Fe	Ti		TiO₂

The division into "arc," "spark," and "superspark" is clearly shown by the table. Maxima of the lines which are used as criteria of class are marked with an asterisk.

range in mean density. Secondly, the basis of the classification has been shown to be the degree of thermal ionization.[12] Granted that the value of the partial electron pressure is low enough, in dwarfs as well as in giants, for thermal ionization to predominate over ionization by collision, a mass of gas will pass through the same *succession* of ionization-stages with changing temperature, whatever the surface gravity. Any given stage of ionization will, however, be reached at a lower temperature, the lower the pressure, since, as pointed out in Chapter X,[13] lowered pressure tends to increase the degree of ionization, and will help to produce a given degree of ionization at a lower temperature.

The Draper system takes no direct account of temperature. It classifies purely by degree of ionization, and therefore, as it relates to atmospheres in which the surface gravities differ widely, it will produce classes that are not homogeneous in temperature; dwarfs will be hotter than giants of the same spectral class. Fowler and Milne [14] anticipated a difference of from 10 to 20 per cent, and differences in this sense and of this order actually occur.[15] Physically it seems to be more important to class together stars having the same atmospheric properties than stars at exactly the same effective temperature, although the latter might conceivably be better suited to some purposes.

Although giant and dwarf stars may be found with very similar spectra, it is well known that they display important differences for individual lines, and these differences have formed the basis for the estimation of spectroscopic parallaxes.[16] If the spectrum of a giant star is compared with the spectrum of a dwarf *of the same temperature*, the two will be found to differ. The line-intensities in the spectrum of the dwarf will place it in a spectral class nearer to the red end of the sequence — if the giant is of Class F5, the dwarf may be a G0 star. There are two ways in which the stars might be brought into the same spectral class; by an alteration of temperature or by an alteration of pressure.

---

[12] See above, p. 193.
[13] P. 141.
[14] M. N. R. A. S., **83**, 403, 1923.
[15] Chapter II, p. 31.
[16] Adams and Joy; Mt. W. Contr. 142, 1917.

If the temperature of the dwarf star were raised, the resulting changes in ionization in its atmosphere would produce changes in the intensities of the lines in the spectrum. At some tempera-

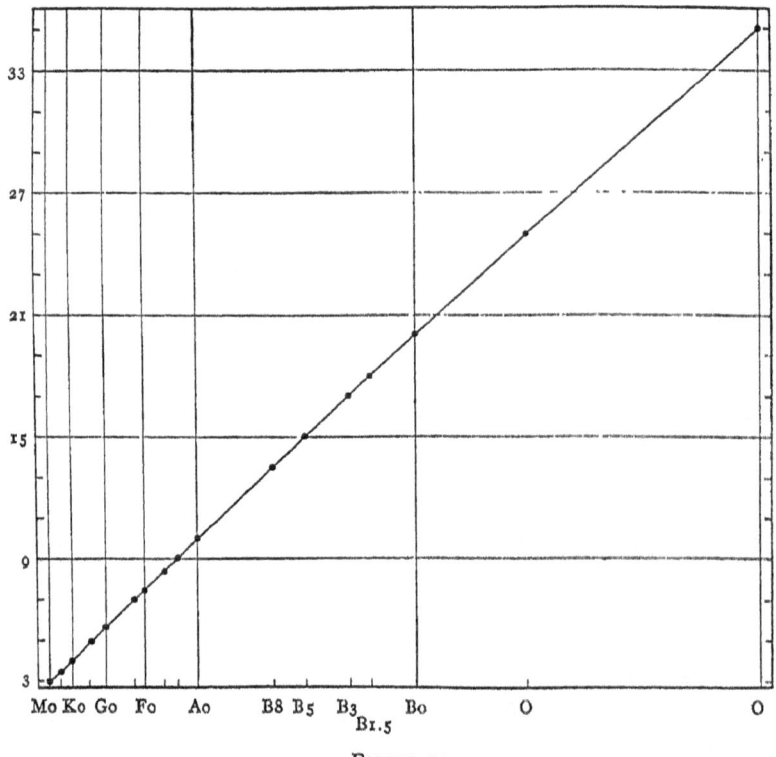

FIGURE 10

Schematic representation of the ionization temperature scale for the sequence of stellar classes. Ordinates are absolute temperatures in thousands of degrees; abscissae are Draper classes. The points representing the different classes have been made to lie on a straight line, so that the temperature range of the corresponding classes shall appear along the axis of abscissae. Vertical lines are drawn through Mo, Ko, Go, Fo, Ao, Bo, and the upper limit of the O class, in order to show the increase in temperature range for the hotter classes.

ture, about 15 per cent higher than the original temperature of the dwarf star, it would give a spectrum resembling that of the giant. If the pressure in the atmosphere of the dwarf star were reduced, the resulting increase in the degree of ionization would also produce changes in the spectral lines, until it gave a spectrum similar to that of the giant. There is, however, no reason to suppose that the changes produced in the intensities of *individual lines* by these temperature and pressure changes would be in all cases exactly equal, although they would in general operate in the same direction. Excitation and ionization conditions

TABLE XXXI

Class	Effective Temperature	Absolute Magnitude		Galactic Concentration		Percent in H.D.C.	Space Number
		d	g	7.0–8.25	]7.0		
B0	20,000°						
to			−0.50			3.52	4.4
B5	15,000						
B8					16	0.96	
B9				50.	9.2	4.70	
A0	11,200			12.8	3.5	10.41	250
A2			+1.50	5.3	1.8	8.89	
A3				2.8	1.2		
A5	8,600		+2.20	1.9	1.9	2.31	
F0	7,000	+2.5		1.6	1.8	5.48	680
F2		+2.9		1.2	1.4	3.37	
F5	6,080	+3.5		1.3	1.0	5.98	
F8		+4.2				4.28	7600
G0	5,460	+4.5	−1.5			4.78	
G5	4,820	+4.8	+0.6			8.98	
K0	4,240	+6.20	+1.05			19.65	160 (giant)
K2						6.85	
K5	3,600	+7.20	+0.50			4.80	22 (giant)
M1	3,380	+10.20	+0.40			2.10	

differ so widely for different atoms that it would be expected that two factors, one of which encourages ionization, while the other discourages recombination, would not in every case balance exactly, even when their mean effect was constant, as it is for any one Draper class.

The Henry Draper Catalogue, as we have emphasized, was made on the basis of the general resemblance of the spectra, an arrangement which corresponds to the greatest physical homogeneity that can be obtained. As regards features of their spectra, it is therefore to be expected that the members of any one class will correspond closely, and care must be exercised in eliminating redundancies from discussions of the homogeneity of the individual classes.

There are, however, other types of discussion, independent of spectroscopic data, and such investigations have shown that the Draper classes have indeed a significance far beyond the mere formation of homogeneous groups of spectra. In illustration of the profound statistical significance of the classification, the table on page 197 of the present chapter contains a brief synopsis of some of the most salient features that have been correlated with spectral class. Successive columns contain the class, the effective temperature,[17] the mean absolute magnitude,[18] the galactic concentration,[19] the percentage of the class in the Draper catalogue,[20] and the computed number per million cubic parsecs.[21]

[17] A. N., 219, 361, 1923. [18] Lundmark, Pub. A. S. P., 34, 147, 1922.
[19] Shapley, H. B. 796, 1924.
[20] Shapley and Cannon, Proc. Am. Ac. Sci., 59, 217, 1924.
[21] Shapley and Cannon, ibid., 59, 230, 1924.

CHAPTER XV

ON THE FUTURE OF THE PROBLEM

THE future of a subject is the product of its past, and the hopes of astrophysics should be implicit in what the science has already achieved. Astrophysics is a young science, however, and is still, to some extent, in a position of choosing its route; it is very much to be desired that present effort should so be directed that the chosen path may lead in a permanently productive direction. The direction in which progress lies will depend on the material available, on the development of theory, and on the trend of thought.

The material already at hand is far from exhaustively analyzed, and it is perhaps premature to contemplate collecting more. But as a science progresses it is often possible to direct the way "by showing the kind of data which it is especially important to improve," and particularly is this the case for astrophysics. In the improvement of the old data, by far the most important requirement is some method of standardizing the intensities of spectrum lines, and of measuring their width, energy distribution, and central intensity. This involves a very difficult and necessary piece of photographic photometry. The problem is an old one that has defied attack for a long time past. It is none the less urgent, and until the attack has been successfully made, many questions, such as are discussed in Chapter III, and other questions, which, for lack of data, we have not been able to discuss at all, must await their precise answers.

Much patient labor, on types of investigation that have already been well worked, still remains to be done. The identification of lines in the spectra of the sun and stars must necessarily be of a laborious nature, but the fact that more than two thirds of the lines in Rowland's table are still unidentified shows how necessary and how large a piece of work this is. One of the things that would greatly assist progress would be a revision of Rowland's table in the light of the recent analysis of the arc and spark spectra of the metals, insertion of the series relations, when known, and the reduction of the wave-length system to International Angstroms.

Another line of work, which lies upon the borderland between astrophysics and pure physics, is the analysis of spectral series. For most of the astrophysically important lines, series relations are already known, but some of the more difficult spectra, such as the spectrum of nitrogen, remain unanalyzed. The analysis of all such spectra is necessary to the advance of astrophysics.

The investigation of stellar spectra has been confined, for the most part, to the region lying between 3900 and 5000, although work on special stars has been carried into the red and the ultra-violet. The use of special dyes should permit work to be carried to about 7900 in the red, and a wave-length of 3500 appears to be accessible in the ultra-violet. There appears to be a large field for an extension of the analysis of stellar spectra into regions of the spectrum that are comparatively unexplored, and the writer hopes in the immediate future to undertake work in this direction.

The types of investigation hitherto mentioned are amplifications of work already in progress. New fields are not easy to predict, but they may be suggested by examining the extent to which present investigation is covering the possibilities of the data. The line *position* and *intensity* data are in full use at the present time. The *form* and *energy distribution* of individual lines, and the study of *asymmetries*, are among the urgent future problems. The measurement of the *polarization* of the light received from the stars has enormous possibilities, but so far very little success has attended such attempts.

The future progress of theory is a harder subject for prediction than the future progress of observation. But one thing is certain: observation must make the way for theory, and only if it does so can the science have its greatest productivity.

ON THE FUTURE OF THE PROBLEM

Observational astrophysics is so vigorous a science that the progress of theory is almost completely determined by the progress of observation.

The most important of the three factors contemplated at the opening of the chapter is perhaps the trend of thought. It is owing to the tendency towards laying stress on observation, and to the general lessening of the distrust of large dimensions, that astrophysics has become possible as a science. The surprising growth of the subject during the last forty years is in great measure the result of this happy chance. The growth of the subject during the next forty years will depend on the coming trend of thought.

The prospect appears encouraging. At the present time the tendency is towards mutual toleration of point of view and to understanding of limitations among the sciences, and a consequent increase of correlation. If the breadth of conception thus engendered develops in the future as it has done in the immediate past, there is hope that the high promise of astrophysics may be brought to fruition.

APPENDICES

I. INDEX TO DEFINITIONS

An attempt has been made to define specifically, at some point in the text, most of the technical terms that are associated with the theory of ionization. For convenience of reference, the most important of these terms are collected into the brief index which is given below. The references are to the pages on which the term is defined.

Atomic life	21,110	Photosphere	35,47
Azimuthal quantum number	8,204	Quantum number	8,204
Boundary temperature	27	Quantum relation	11
Displacement Rule	13	Residual intensity	51
Effective level	135	Reversing layer	47,49
Effective temperature	27	Rydberg constant	14,155
Excitation potential	15	Saturation	52,135
Fractional concentration	105	Series notation	55,203
Inner quantum number	204	Spectroscopic valency	10
Ionization potential	15	Subordinate lines	12,100
Ionization temperature	30,132	Temperature class	24,112
Marginal appearance	105,135,179	Total quantum number	8,205
Optical depth	27, 35	Ultimate lines	11,111
Partial electron pressure	109	Valency	10
Partition function	107	Wings	50,179

II. SERIES RELATIONS IN LINE SPECTRA

A synopsis of the normal series relations in line spectra has been published by Russell and Saunders (Ap. J., **61**, 39, 1925). A transcription of the passages containing definitions of spectroscopic quantities that are mentioned in the present volume is given below:

"Every spectral line is now believed to be emitted (or absorbed) in connection with the transition of an atom (or molecule) between two definite (quantized) states, of different energy-content — the frequency of the radiation being exactly proportional to the change of energy. The wave-number of the line may therefore be expressed as the difference of two *spectroscopic terms* which measure, in suitable units, the energies of the initial and final states. Combinations between these terms occur according to definite laws, which enable us to classify them into systems, each containing a number of series of terms, which are usually multiple. . . .

"Any term y may be expressed in the form $y = R/(m + x)^2$, where R is the Rydberg constant and m an integer. For homologous components of successive terms of the same series, m changes by unity, while the "residual" x is sometimes practically constant (Rydberg's formula), or, more often, is expressible in the form $\mu + a/m$ (Hicks's formula), or $\mu + ay$ (Ritz's formula). In many cases this approximation fails for the smaller values of m; and prediction becomes very uncertain, though a plot of the residuals usually gives a smooth curve. . . .

"The *principles of selection*, which determine what combinations among these numerous terms give rise to observable lines, are very simply expressed in terms of two sets of quantum numbers.

"The *azimuthal quantum number* (k) is 1 for all terms of the s-series, 2 for those of the p-series, 3 for the d's, 4 for the f's, 5 for the g's, 6 for the h's, and so on.

"Combinations usually occur only between terms of *adjacent series* for which the values of k *differ by a unit*. A great many lines are, however, known for which the change of k is 0, and a few for which it is 2. In the simpler spectra, such lines are faint, except when produced under the influence of a strong magnetic field; but in the more complex spectra they are often numerous and strong.

"The *inner quantum number* (j) differs from one component of a multiple term to another, and also in the various series and systems, according to the following scheme.

k	Series	Singlets	Doublets	Triplets	Quartets	Quintets	Sextets	Septets
1	s	$j=0$	1	1	2	2	3	3
2	p	1	1,2	0,1,2	1,2,3	1,2,3	2,3,4	2,3,4
3	d	2	2,3	1,2,3	1,2,3,4	0,1,2,3,4	1,2,3,4,5	1,2,3,4,5
4	f	3	3,4	2,3,4	2,3,4,5	1,2,3,4,5	1,2,3,4,5,6	0,1,2,3,4,5,6
5	g	4	4,5	3,4,5	3,4,5,6	2,3,4,5,6	2,3,4,5,6,7	1,2,3,4,5,6,7

"Combinations occur only between terms for which j differs by 0 or ± 1. If, however, $j = 0$ in both cases, no radiation occurs. Lines corresponding to a change of $j = \pm 2$ are found in strong magnetic fields, and a very few in their absence.

APPENDIX   205

"The combination of two multiple terms gives rise, therefore, to a group of lines (which may number as many as eighteen). Such groups have been called *multiplets* by Catalan. Their discovery has afforded the key to the many-lined spectra. ...

"In such a group, those lines for which the changes in j and k, in passing from one term to the other, are of the same sign, are the strongest, and those in which they are of opposite sign the weakest. These intensity relations are of great assistance in picking out the multiplets.

"Combinations between terms of different systems (consistent with the foregoing rules) often occur. Such lines are usually, though not always, faint. ...

"The serial number m of the term (which is equivalent to the *total-quantum number*) plays quite a subordinate rôle, being of importance only when series formulae have to be calculated. An extensive analysis of a spectrum is possible without it, though determination of the limits of the series, and the ionization potential, demands its introduction."

III. MATERIAL USED IN CHAPTER VIII

THE line intensities quoted in Chapter VIII were derived from the spectra of the stars enumerated below in Table XXXII. Successive columns contain the Draper class, the name of the star, the Boss number, the visual apparent magnitude, and the reduced proper motion H. The stars within each class are arranged in order of right ascension.

TABLE XXXII

Class	Star	Boss	m	H	Class	Star	Boss	m	H
B9	λ Cen	3054	3.3	1.7	A2	q Vel	2723	4.1	5.1
A0	η Phe	148	4.5	−0.5		ʃ Sgr	4832	2.7	0.0
	ν For	474	4.7	0.1		ε Gru	5880	3.7	4.1
	s Eri	611	4.5	4.3	A3	τ₃ Eri	696	4.2	5.1
	α Dor	1081	3.5	2.1		β Pic	1446	3.9	3.7
	δ Vel	2356	2.0	1.9		λ Mus	3092	3.8	3.9
	β Car	2493	1.8	3.2		β Pav	5315	3.6	2.1
	γ Cen	3302	2.4	3.9		α PsA	5916	1.3	4.1
	γ TrA	3879	3.1	2.2	F0	α Hyi	458	3.0	5.0
	ε Sgr	4645	2.0	2.7		α Car	1622	0.9	−4.6

Class	Star	Boss	m	H	Class	Star	Boss	m	H
F0	ι Car	2503	2.2	−0.7	K0	ε Crv	3172	3.2	2.3
	γ Boo	3722	3.0	4.3		π² Hyi	3622	5.5	4.6
	α Cir	3739	3.4	5.0		θ Cen	3623	2.3	6.7
	β TrA	4030	3.0	6.2		ζ Lup	3864	3.5	4.0
	θ Sco	4457	2.0	−2.6		ν Lib	3962	5.3	−1.2
F2	η Sco	4361	3.4	5.7		γ Apo	4168	3.9	4.6
	π Sgr	4874	3.0	1.0		ε Sco	4272	2.4	6.5
F5	δ Vol	1917	4.0	0.6		γ Sgr	4568	3.1	4.6
	α CMi	2008	0.5	6.1		δ Sgr	4628	2.8	1.3
	ρ Pup	2153	2.9	2.9		λ Sgr	4665	2.9	4.4
	b Vel	2324	4.1	0.0		ξ² Sgr	4809	3.6	1.4
	d Oph	4421	4.4	5.4		τ Sgr	4857	3.4	5.5
	ι₁ Sco	4492	3.1	−3.9		α Ind	5281	3.2	2.5
	κ Pav	4778	var.	0.0		c₂ Aqr	5963	3.8	0.5
F8	ζ Tuc	55	4.3	10.9		ι Gru	5965	4.1	4.9
	α For	723	4.0	8.3	K2	ε Cru	3218	3.6	5.1
	γ Lep	1420	3.8	9.7,7.1		δ Mus	3377	3.6	5.7
	δ CMa	1839	2.0	−4.5		α TrA	4250	1.9	−0.6
G0	β Hyi	74	2.9	9.7		β Ara	4406	2.8	0.6
	α Aur	1246	0.2	3.4		α Tuc	5747	2.9	2.6
	β Lep	1323	3.0	2.9	K5	α Tau	1077	1.1	2.6
	ξ Pup	2065	3.5	−2.3		π Pup	1896	2.7	−2.8
	ε Leo	2618	3.1	1.5		σ Pup	1972	3.3	4.7
	l Car	2628	var.	1.1		q Car	2739	3.4	1.7
	ζ Cap	5507	3.9	(1.1)		α Apo	3746	3.8	1.4
G5	α Ret	994	3.4	2.6		η Ara	4265	3.7	2.5
	μ Vel	2875	2.8	2.3		ζ² Sco	4292	3.8	5.9
	ξ Hya	3042	3.7	5.3		ζ Ara	4304	3.1	1.5
	β Crv	3280	2.8	1.7	Ma	β And	259	2.4	−0.8
	γ Hya	3449	3.3	2.9		γ Hyi	899	3.2	−1.3
	β CrA	4871	4.2	2.0		α Ori	1468	0.9	−0.0
	δ Pav	5138	3.6	9.7		α Sco	4193	1.2	−2.5
K0	α Phe	78	2.4	5.6	Mb	τ₄ Eri	759	4.0	3.2
	α Cas	135	2.5	1.4		γ Cru	3263	1.6	3.8
	β Phe	245	3.4	1.6		σ Lib	3837	3.4	3.3
	δ Phe	336	4.0	5.4		α Her	4373	3.5	0.9
	β Ret	875	3.8	6.2		η Sgr	4617	3.2	4.9
	δ Lep	1456	3.9	8.1		β Gru	5854	2.2	2.8
	β Col	1459	3.2	6.2					

IV. INTENSITY CHANGES OF LINES WITH UNKNOWN SERIES RELATIONS

The following tabulation shows the intensity changes of lines of unknown series relations that occur in the hotter stars. The arrangement follows that of Table XIX. Notes on the maxima and blends are appended.

TABLE XXXIII

Atom	Line	ζPup	τCMa	29CMa	Bo	B0	B1	B2	B3	B5	B8	B9	A0	A2	Note
C++	4649	0.0	2.0	9.0	6.0	0.0	0.0	0.0	0.0	0.0	0.0	0.0	0.0	..	1
N+	3996.9	0.0	0.0	0.0	0.0	5.0	6.0	7.0	..	9.0	..	..	5.0	0.0	1
N++	4515.0	..	9.0	4.0	4.0	1.0	0.0	0.0	0.0	0.0	0.0	0.0	0.0	0.0	2
	4097.5	..	15.0	8.0	5.0	..	0.0	0.0	0.0	0.0	0.0	0.0	0.0	0.0	3
O+	4943.4	0.0	0.0	0.0	0.0	4.0	4.0	3.0	0.0	0.0	0.0	0.0	0.0	0.0	1
	4941.2														
	4705.3	0.0	0.0	0.0	2.0	4.0	4.0	3.0	0.0	0.0	0.0	0.0	0.0	0.0	2
	4699.2														
	4676.2	0.0	0.0	0.0	2.0	5.0	..	7.0	0.0	0.0	0.0	0.0	0.0	0.0	3
	4661.6	0.0	0.0	0.0	2.0	5.0	..	7.0	0.0	0.0	0.0	0.0	0.0	0.0	4
	4649.1	0.0	0.0	0.0	6.0	9.0	12.0	9.0	..	4.0	0.0	0.0	0.0	0.0	5
	4641.8	0.0	0.0	0.0	0.0	3.0	10.0	7.0	..	0.0	0.0	0.0	0.0	0.0	6
	4596.2	0.0	0.0	0.0	1.0	5.0	..	6.0	..	3.0	0.0	0.0	0.0	0.0	7
	4591.0	0.0	0.0	0.0	1.0	5.0	..	6.0	..	3.0	0.0	0.0	0.0	0.0	8
	4417.0	0.0	0.0	0.0	5.0	6.0	11.0	3.0	..	2.0	0.0	0.0	0.0	0.0	9
	4415.9	0.0	0.0	0.0				3.0	..	2.0	0.0	0.0	0.0	0.0	10
	4366.9	0.0	4.0	4.0	4.0	..	6.0	6.0	..	1.0	0.0	0.0	0.0	0.0	11
	4075.9	0.0	3.0	0.0	2.0	6.0	8.0	6.0	0.0	0.0	0.0	0.0	0.0	0.0	12
	4072.2	0.0	0.0	0.0	2.0	7.0	9.0	6.0	0.0	0.0	0.0	0.0	0.0	0.0	13
	4069.9	0.0	0.0	0.0	4.0	6.0	8.0	6.0	0.0	0.0	0.0	0.0	0.0	0.0	14
S+	4815	0.0	0.0	0.0	0.0	0.0	0.0	0.0	0.0	0.0	x	0.0	0.0	0.0	1
	4174.5	0.0	0.0	0.0	0.0	0.0	0.0	0.0	0.0	0.0	3.0	0.0	0.0	0.0	2
	4162.9	0.0	0.0	0.0	0.0	0.0	0.0	0.0	0.0	0.0	3.0	0.0	0.0	0.0	3
S++	4295	0.0	0.0	0.0	0.0	0.0	1.0	0.0	0.0	0.0	0.0	0.0	0.0	0.0	4
	4285.1	0.0	0.0	0.0	0.0	0.0	6.0	4.0	0.0	0.0	0.0	0.0	0.0	0.0	5
	4253.8	0.0	0.0	0.0	0.0	6.2	6.8	6.0	0.0	0.0	0.0	0.0	0.0	0.0	6

208 APPENDIX

Notes to Table XXXIII

Atom	Note	Maximum	Remarks
$C++$	1	29 CMa	Line blended, in stars cooler than B0, with $V+$ 4649.1. Attributed by Fowler and Milne, and by Hartree, to $C+++$
$N+$	1	B5	Unblended
$N++$	2	τ CMa	
	3	τ CMa	Blended with the $Si+++$ line at 4096, which is probably effective throughout the whole range
$O+$	1	B0–B1	
	2	B0–B1	
	3	B2?	
	4	B2?	
	5	B1	Blended with $C++$ line at 4649, which preponderates in stars hotter than B0, and probably contributes largely in that class
	6	B1	
	7	B2	
	8	B2	
	9	B1	
	10	B1	
	11	B1–B2?	Certainly another line is here involved, but it has not been identified
	12	B1	
	13	B1	
	14	B1	
$S+$	1–3	B8	Lines recorded by Lockyer; not measured by the writer
$S++$	4	B1	Line recorded by Lockyer. Intensity from H.A., 28; not measured by the writer
	5	B1	
	6	B1	Recorded by H. H. Plaskett in 10 Lacertae

V. MATERIAL BEARING ON THE CLASSIFICATION OF A STARS, QUOTED IN CHAPTER XII

In illustration of the problem of Class A, observations of sixty-two stars are collected in the following table. Successive columns contain the H.D. number, the name of the star, the apparent magnitude, the reduced proper motion H, and the spectral class. Then follow columns which indicate the presence (x) or absence of metallic lines, the quality of the lines (sharp lines being represented by the letter s and hazy lines by the letter h), the presence of wings to the hydrogen lines, and the strength of the $Sr+$ line at 4077 and the $Si+$ lines at 4128, 4131.

APPENDIX 209

The stars in each class are arranged in order of increasing strength of metallic lines, and it will be seen that this feature is correlated with the strength of the silicon and strontium lines, but not with the line quality or the hydrogen wings, nor with the reduced proper motion.

TABLE XXXIV

H. D.	Star	m	H	Class	Metallic Lines	Line Quality	H wings	Sr+	Si+
120198	84 UMa	5.53	6.4	Aop	x			9	10
108662	17 Com	5.38	2.7	Aop	x			7	8
170397	Br 2314	5.99	2.7	Aop	x	h		6	9
133029	+47°2192	6.16	—	Aop	x	h		5	11
140160	χ Ser	5.26	3.5	Aop	x			10	5
94334	ω UMa	4.34	3.6	Ao	x	s		2	3
58142	21 Lyn	4.45	2.9	Ao		h		—	4
192913	+27°3668	6.69	—	Aop		h		?	7
225132	2 Cet	4.62	1.2	Ao		h		—	4
41841	89 Lep	5.50	2.2	Ao				—	—
222661	ω² Aqr	4.62	4.8	Ao		h		—	—
87887	15 Sex	4.6	1.9	Ao				—	5
213323	38 Peg	5.51	3.3	Ao				—	—
25642	λ Per	4.33	2.2	Ao				—	4
114330	θ Vir	4.44	3.2	Ao		s		—	3
109485	23 Com	4.78	4.1	Ao		s		—	3
103632	η Cra	5.16	3.8	Ao			x	—	—
110411	ρ Vir	4.95	5.6	Ao			x	—	—
133962	k Boo	5.59	4.6	Ao			x	—	—
188260	13 Vul	4.50	2.5	Ao		h	x	—	4
124224	Pi 12	4.90	3.8	Aop				—	11
183056	4 Cyg	5.2	−0.4	Aop				—	9
183986	+35°3658	6.04	—	Ao			x	—	5
196502	73 Dra	5.18	1.2	A2	x	s		12	10
148367	υ Oph	4.68	4.3	A2	x	s	x	9	8
118022	78 Vir	4.93	3.5	A2p	x			10	9
182564	π Dra	4.63	2.9	A2	x			7	7
125337	λ Vir	4.60	−1.9	A2	x	h		7	6
214734	30 Cep	5.21	1.6	A2	x			3?	5
7804	89 Psc	5.28	4.2	A2	x		x	5	7
220825	κ Psc	4.94	—	A2p	x			6	7
72968	3 Hya	5.61	2.8	A2p	x			6	7
56405	Paris 8971	5.39	4.7	A2	x	s	x	3?	6
20677	32 Per	4.98	3.8	A2	x		x	—	5

APPENDIX

H. D.	Star	m	H	Class	Metallic Lines	Line Quality	H wings	Sr+	Si+
48250	12 Lyn	4.89	1.8	A2	x		x	—	5
107612	—Com	6.56	—	A2p			x	9	—
18519,20	ε Ari	4.6	1.1	A2,A2		s		3	3
107966	13 Com	5.10	2.7	A2				—	—
108382	16 Com	5.04	0.6	A2				—	—
108945	21 Com	5.39	1.9	A3p	x			11	9
108642	+26°2138	6.48	—	A3	x			8	6
89904	27 LMi	6.1	3.0	A3	x	h	x	6	9
108651	+26°2353	6.69	—	A3	x	h		7	6
170296	γ Scu	4.73	−0.5		x	h	x	—	7
115331	196 Cen	6.0	3.4	A3p		h		9	—
108486	+26°2352	6.57	—	A3				7	—
104321	π Vir	4.57	2.2	A3				5	—
222345	ω₁ Aqr	5.16	4.6	A5	x			—	9
14690	70 Cet	5.62	4.3	A5	x			7	?
189849	15 Vul	4.74	3.3	A5	x	s	x	9	6
28546	81 Tau	5.49	5.7	A5	x			9	9
40536	2 Mon	5.10	4.0	A5	x	h?		7	7
15089	ι Cas	4.59	0.5	A5p	x			12	9
91312	Gr 1658	4.85	5.6	A5	x	s	x	5	6
159560	ν² Dra	4.95	6.0	A5				8	—
90277	30 LMi	4.85	5.0	F0	x			9	9
57749	Pi 86	5.83	2.3	F0	x			9	6
92787	Pi 135	5.28	7.6	F0	x	s		9	6
112429	8 Dra	5.27	3.0	F0	x			7	7
28485	80 Tau	5.70	5.8	F0		h		6	5
28677	85 Tau	6.04	—	F0	—			7	—

SUBJECT INDEX

A, Class............................166
   classification..................168
   metallic lines and band absorption....................167
   peculiar........................172
   observational material........208
Abnormal abundance, effect on spectrum.........................172
Abnormal conditions, effect on spectrum.........................171
Absent elements................... 86
Absolute magnitudes for Draper classes..........................197
Absorbing efficiencies of atoms....136
Absorption coefficient............110
Absorption lines...............11, 50
Abundance of atomic species, 135, 177, 183
Abundance of atoms, terrestrial.... 5
Aluminum.....................68, 122
Arc spectrum...................... 13
Atomic life....................21, 110
   life, astrophysical..........22, 158
   nucleus......................... 4
   number.......................... 5
   states, probabilities of........ 23
   weights......................... 5
Azimuthal quantum number....8, 204
Balmer lines in A stars...........166
   lines, maximum................166
   series limit.................... 42
Barium........................85, 126
Blackness of continuous background 49
Blending of lines.................171
Bohr's Table...................... 9
Boundary temperature............. 27
Calcium.......................70, 122
Carbon.....................59, 121, 207
   compounds..................... 61
   spectroscopic and chemical valencies......................... 11
Central intensities of spectrum lines. 51
Central temperature.............. 27
Chemical symbols................. 5

Chemical valency................. 10
Chromium.....................77, 124
Chromosphere.............35, 47, 159
Classification, meaning of........190
Classification, principles of......191
Cobalt........................... 80
Color indices of stars............ 49
Conductivities of flames..........112
Consistency of temperature scale...130
Continuous background of stellar spectrum......................46, 48
Copper........................... 80
Critical potential................ 14
Critical potential, astrophysical determination.....................156
c-stars..........................173
Definitions, index to.............203
Density and radius of second type stars............................ 36
Displacement rule.............13, 20
Draper classes, homogeneity......193
Draper system....................191
Duration of atomic states........ 21
Earth, composition of crust......185
   origin of.......................185
   total, composition of..........187
Effective level................28, 136
Effective temperature............ 27
Effective temperature, Draper classes..........................197
Effect of conditions on spectrum.... 24
Electrons, extra-nuclear.......... 8
Electron transfer................. 11
Elements......................... 5
Elements, in stellar atmospheres... 55
Emission lines, production of....11, 53
Emission lines, stars showing..... 53
Energy distribution in solar spectrum 49
Energy distribution in stellar spectrum........................... 48
Equivalent outer orbits, electronic.. 10
Even number rule.................177
Excitation....................... 11
Excitation potential.............. 15

SUBJECT INDEX

Flash spectrum..............40, 53
Formulae in theory of ionization....106
Fractional concentration..........105
Furnace experiments.............112
Galactic concentrations of Draper classes......................197
Gallium........................81
Gaseous nebulae, continuous spectrum........................49
Giant and dwarf spectra..........195
Giant and dwarf temperature differences........................31
Helium....................58, 121
Hot spark......................18
Hydrogen, atom................12
    continuous spectrum........57
    energy levels...............13
    line intensities.............121
    peculiar behavior...........56
    secondary spectrum.........56
    stellar spectra..............55
Inner quantum number..........204
Ionization................11, 15, 97
    as criterion of pressure.......44
    potential....................15
    " and atomic number.20
    " astrophysical......156
    " and temperature of maximum........157
    temperature................30
    temperature scale....33, 132, 133
Ionized atom, lines of...........101
Intensity of lines.............23, 116
    accuracy..................118
    estimation.................117
    observational material......116
Intensity scales................117
Interior of star, ionization in......18
Iron.........................79, 125
Isotopes........................5
Law of Mass Action.........105, 110
Lead..........................86
Level of origin of lines..........134
Life of atom....................21
Life of atom, astrophysical....22, 158
Limit of Balmer series...........42
Line intensity, data on..........121
    notes on..................127
    unknown lines.............207
Lines of unknown origin.........55

Lithium........................59
Low temperature conditions, stellar atmosphere...................94
Magnesium.................67, 122
Magnesium, compounds..........68
Manganese..................78, 124
Marginal appearance....105, 135, 179
Marginal appearance, observational data........................181
Mass numbers of isotopes.........5
Maximum of lines..............106
Meteorites, composition.........187
Molecule, ionization of..........19
Molybdenum...................84
Nickel.........................80
Niobium.......................84
Nitrogen.....................63, 207
Nucleus, atomic.................4
O, Class......................162
Occurrence of elements in stars....5
Optical depth...............27, 35
Origin of line spectra............11
Orion nebula, continuous spectrum.49
Oxygen.....................65, 207
Oxygen, compounds..............66
Palladium.....................84
Partial electron pressure.........109
Partition function..............107
Percentage of stars in Draper classes 197
Photosphere.................35, 47
Photoelectric ionization.........159
Physical constants, astrophysical evaluation...................155
Physical constants, required by ionization theory...............108
Potassium.....................70
Pressure, atmospheres of stars...25, 34
    from Balmer series limit......44
    from flash spectrum..........40
    from ionization theory........45
    from line sharpness..........38
    from line width..............39
    from radiative equilibrium....40
    shift of lines..............26, 38
    summary....................45
    gradient in outer layers of star 34
Quantum number................8
Quantum relation...............11

SUBJECT INDEX 213

Radiative equilibrium in outer layers of star.......... 40
Radium.......... 86
Radius and density of second type stars.......... 36
Rare earths.......... 85
Relative abundance of atoms, estimation.......... 183
   abundance of atoms in stellar atmosphere.......... 177
   intensities of spectrum lines.. 23
Reorganization time, *see* atomic life
Residual intensity.......... 51
Reversing layer.......... 50
Reversing layer, mass and dimensions.......... 47
Rhodium.......... 84
Rubidium.......... 81
Ruthenium.......... 84
Rydberg constant.......... 14, 155, 204
Saturation.......... 52, 135
Scandium.......... 72, 123
Series notation.......... 55
Silicon.......... 24, 68, 122
Silicon stars.......... 169
Silver.......... 84
Sodium.......... 67
Sodium atom.......... 8
Solar atmosphere.......... 47
   energy distribution.......... 49
   intensities.......... 113
Space numbers for Draper classes... 197
Spark spectrum.......... 14
Special problems in stellar atmospheres.......... 161
Spectroscopic valency electrons.... 10
Stark effect.......... 26
Stars used for intensity estimates.. 119
Stellar atmosphere compared with earth's crust.......... 184

Stellar reversing layer.......... 91
Strontium.......... 81, 126
Strontium stars.......... 169
Structure of absorption line.......... 180
Subordinate lines.......... 12
Subordinate series.......... 99
Sulphur.......... 70, 207
Surface gravity.......... 35, 36
Symmetry number, *see* spectroscopic valency electrons

Temperature class.......... 24, 112
   scale.......... 28, 33
   scale, ionization.......... 33, 133
   bright stars.......... 31
   giant and dwarf.......... 31
Theory of solution.......... 110
Tin.......... 85
Titanium.......... 72, 123
Titanium compounds.......... 75
Total pressure in stellar atmospheres 110
Total quantum number.......... 8
Typical giant star.......... 41

Ultimate lines.......... 11, 94, 111
Uniformity of stellar atmospheres.. 178

Valency, chemical.......... 10
Valency, spectroscopic.......... 10
Vanadium.......... 75, 124

Weights of atomic states.......... 107
Width of lines.......... 39
Wings.......... 51
Wolf-Rayet stars.......... 164
   absorption in.......... 164
   continuous spectrum.......... 49

Yttrium.......... 82, 129

Zeemann effect.......... 25
Zinc.......... 80, 126
Zirconium.......... 83

NAME INDEX

Abbot............30, 48, 51
Adams......54, 75, 140, 149, 192, 195
Anderson..............26
Aston..................5
Babcock............25, 38
Baillaud..............48
Baldet................62
Baxandall....55, 63, 64, 149, 172, 173
Belopolsky............173
Bohr............8, 42, 108
Bottlinger............51
Bowen................18
Brandt................17
Brill.................29
Brooks................68
Brooksbank............66
Burns................159
Butler................63
Campbell, W. W......60, 163
Cannon, A. J....54, 163, 190, 108
Catalan..........17, 77, 84
Chenault..............17
Clarke, F. W....5, 184, 185, 187
Coblentz..............30
Compton, K. T....17, 56, 59
Cortie................66
Coster................23
Curtis...............159
Curtiss...............58
Davies.............17, 19
de Forcrand...........60
de Gramont........59, 111
Dempster.............22
Deslandres............58
Dorgelo..............23
Duffendack...........19
Dyson................86
Eddington....27, 34, 41, 53, 180
Einstein........23, 113, 165
Eldridge.............17
Evershed......26, 38, 58, 62
Fairfield......56, 57, 159
Fermi................23

Foote................17
Fowler, A., 11, 14, 17, 38, 50, 60, 62, 64, 66, 68, 75, 186, 191
Fowler, R. H., 17, 37, 44, 61, 70, 94, 106, 110, 108, 133, 135, 156, 160, 179
Franck..........19, 22, 23
Füchtbauer...........23
Gale................75
Giebeler.............86
Gieseler.............17
Gousmid..............86
Grotrian...........17, 22
Guckel..............108
Hale..............25, 75
Harper......70, 117, 176, 194
Harrison............181
Hartley..............81
Hartmann.............75
Hartree...........17, 61
Heger.............67, 74
Hertzsprung......31, 175
Hoffmann.............23
Hopfield.........17, 186
Horton............17, 19
Howe................44
Hubble............49, 57
Huggins..............57
Hulburt..............19
Huthsteiner..........17
Jeffreys............189
Johnson, M. C.....56, 84
Johnson, R. C........62
Joy........54, 140, 192, 195
Kiess, C. C....13, 17, 72, 84, 86, 173
Kiess, H. K..........17
King, A. S....34, 38, 60, 84, 113
King, E. S...........29
Knipping.............19
Kohlschütter....51, 140
Kohn................60
Kossell...........13, 23
Kramers...........8, 23
Krüger..............19

NAME INDEX

Landé...........................25
Lee............................71
Lindblad...........48, 57, 62, 169
Lindemann....................26
Lockyer, J. N........64, 70, 172, 173
Lundmark......................198
Lunt........................5, 85
Luyten.............67, 162, 170
Lyman................18, 57, 58
Maury.......................173
McLennan..................18, 63
Meggers...........13, 17, 72, 75, 82
Menzel, 44, 66–68, 71, 74–76, 79–81, 85, 109, 116, 133, 166, 181
Merrill, G. P....................187
Merrill, P. W.................54, 83
Merton 25, 60, 62
Mie...........................22
Millikan........................18
Milne, 17, 22, 28, 36, 37, 44, 48, 50, 61, 70, 93, 97, 106–110, 113, 133, 135, 140, 156, 158, 166, 179
Mitchell....................58, 86
Mohler........................17
Mulliken......................61
Newall........................63
Nicholson, J. W...............43
Noyes.......................112
Pannekoek................35, 140
Paschen.............14, 17, 156
Payne 20, 38, 43, 58, 60, 68, 156, 163, 183
Plaskett, H. H., 14, 30, 48, 51, 54, 59, 60, 64, 65, 85, 156, 163, 184
Plaskett, J. S......60, 71, 162, 163, 171
Pluvinel.......................62
Ramage........................81
Rognley........................17
Rosenberg....................29
Rowland......................127
Ruark.........................17
Rufus......................62, 72
Russell, 17, 26, 37, 39, 40, 44, 47, 52, 55, 56, 59, 70, 79, 80, 110, 175, 178, 184, 203

Saha..............43, 85, 105, 113
St. John...................25, 38
Sampson......................30
Saunders............38, 55, 203
Scheiner.....................29
Schwarzschild.......51, 137, 160
Seares.......................31
Shane........................62
Shapley, 36, 40, 51, 62, 162, 168, 170, 185, 198
Shaver........................17
Shrum........................64
Slipher......................75
Smyth.....................17, 19
Sommerfeld........13, 17, 21, 23
Sponer........................17
Stark.........................25
Stewart 26, 37, 39, 40, 44, 47, 91, 110, 175
Strutt........................66
Takamine......................26
Tate..........................17
Thomas.......................18
Turner........................22
Udden.........................17
Urey........................108
Van Maanen..................162
Vegard........................63
Violle........................60
Von Zeipel..................185
Walters...........13, 72, 77, 79
Washington............5, 184, 185
Webb..........................22
Wien..........................22
Wilsing...................29, 48
Wilson, E. B................162
Wilson, H. A................112
Wilson, H. H................162
Woltjer.....................113
Wood..................22, 42, 57
Wright......43, 56, 57, 66, 163, 173
Young............79, 117, 176, 192
Zeemann......................25